Macromolecular Symposia

Symposium Editor: S. Kobayashi

Editor: I. Meisel
Associate Editor: S. Spiegel
Assistant Editors: H. Beattie, C.S. Kniep

Executive Advisory Board: M. Antonietti, M. Ballauff, H. Höcker,
K. Kremer, H.E.H. Meijer, R. Mülhaupt,
A.D. Schlüter, H.W. Spiess, G. Wegner

157

pp. 1–257

July 2000

Macromolecular Symposia publishes lectures given at international symposia and is issued irregularly, with normally 14 volumes published per year. For each symposium volume, an Editor is appointed. The articles are peer-reviewed. The journal is produced by photo-offset lithography directly from the authors' typescripts.
Further information for authors can be obtained from:
Editorial office "Macromolecular Symposia"
Wiley-VCH
P. O. Box 10 11 61, D-69451 Weinheim,
Germany
or for parcel and courier services: Pappelallee 3, D-69469 Weinheim,
Germany
Tel. +49 (0) 62 01/6 06-2 34 or -2 38; Fax +49 (0) 62 01/6 06-3 09 or 5 10; macromol@wiley-vch.de
http://www.wiley-vch.de/home/macrosymp
Suggestions or proposals for conferences or symposia to be covered in this series should also be sent to the Editorial office at the address above.

Macromolecular Symposia:
Annual subscription rates 2000
Germany, Austria € 1018 (DM 1991,03); Switzerland SFr 1678; other Europe € 1018; outside Europe US $ 1228.
Macromolecular Package, including Macromolecular Chemistry & Physics (18 issues), Macromolecular Rapid Communications (18 issues), Macromolecular Theory & Simulations (9 issues) is also available. Details on request.
Packages including Macromolecular Symposia and Macromolecular Materials & Engineering are also available. Details on request.
Single issues and back copies are available. Please, inquire for prices.

Orders may be placed through your bookseller or directly at the publishers: WILEY-VCH Verlag GmbH, P.O. Box 10 11 61, D-69451 Weinheim, Germany, Tel.: (0 62 01) 6 06–0, Telefax (0 62 01) 60 61 17, Telex 46 55 16 vchwh d. E-mail: subservice@wiley-vch.de

Macromolecular Symposia (ISSN 1022-1360) is published with 14 volumes per year by WILEY-VCH Verlag GmbH, P.O. Box 10 11 61, D-69451 Weinheim, Germany. Air freight and mailing in the USA by Publications Expediting Inc., 200 Meacham Ave., Elmont, NY 11003. Periodicals postage pending at Jamaica, NY 11431. POSTMASTER: send address changes to Macromolecular Symposia, Publications Expediting Inc., 200 Meacham Ave., Elmont, NY 11003.

**International Union
of Pure and Applied Chemistry**
Macromolecular Division

In conjunction with

The Chemical Society of Japan
The Society of Polymer Science, Japan
Japan Chemical Innovation Institute

Invited lectures presented at the

International Symposium
on Ionic Polymerization

Kyoto, Japan
July 19–23, 1999

Symposium Editor

Shiro Kobayashi
Department of Materials Chemistry, Graduate School of Engineering
Kyoto University, Kyoto 606-8501, Japan

Contents of Macromol. Symp. 157

**International Symposium on Ionic Polymerization,
Kyoto, Japan, 1999**

Preface
S. Kobayashi

* The asterisk indicates the name of the author to whom inquiries should be
addressed

Author Index

Preface

The International Symposium on Ionic Polymerization (IP'99) sponsored by The International Union of Pure and Applied Chemistry was held on July 19–23, 1999 in Kyoto, Japan. The symposium was also sponsored by The Chemical Society of Japan, The Society of Polymer Science, Japan, The Society of Synthetic Organic Chemistry, Japan, and Japan Chemical Innovation Institute. This is regarded as the third symposium in the series of The International Symposium on Ionic Polymerization, following up the successful symposia in Istanbul (1995) and in Paris (1997).

The symposium aimed to bring together scientists and engineers from all over the world being interested in ionic polymerizations and related areas and to promote research developments in these fields by an exchange of information and stimulating new ideas. The research area covered in this symposium was directed toward the traditional field of cationic, anionic and ring-opening polymerizations, as well as more broadly to polymer synthesis, including radical polymerization, metal-catalyzed polymerization, polycondensation, enzymatic polymerization, and new polymer architecture.

Approximately 260 active participants from 14 countries attended the symposium. A total of 94 oral reports including invited lectures were presented in two parallel sessions, and 61 posters were also presented. The presentations were of high quality and in the cutting edge of science and technology of these fields. Very vivid and excellent discussions took place during and after the scientific program, definitely contributing to the realization of the purpose of this symposium. This special issue of *Macromolecular Symposia* covers the papers from invited lectures, which will help scientists and engineers to find the future direction in ionic polymerization, as well as in other important fields in polymer synthesis.

The organization of this symposium was made possible with the help and collaboration of all the Committee and Board members. We deeply thank these people for their big effort. Our sincere thanks go to the following corporations for their support: Ajinomoto Co., Inc., Asahi Chemical Industry Co., Ltd., Asahi Glass Co., Daicel Chemical Industries, Ltd., Daikin Industries, Ltd., Denki Kagaku Kogyo Kabushiki Kaisha, DuPont Kabushiki Kaisha, Harima Chemicals, Inc., Hitachi Chemical Co., Ltd., Japan Chemical Innovation Institute, Japan PMC Corporation, JSR Corporation, Kaneka Corporation, Kuraray Co., Ltd., Lion Corporation, Mitsubishi Rayon Co., Ltd., Mitsui Chemicals, Inc., Nippon Zeon Co., Ltd., Nissei Sangyo Co., Ltd., Polyplastics Co., Ltd., Sumitomo Bakelite Co., Ltd., Sumitomo Chemical Co., Ltd., and Toyo Ink MFG Co., Ltd.

Kyoto
October 1999   Shiro Kobayashi

Macromol. Symp. 157, 1–11 (2000)   1

Ring-Opening Polymerization Processes Involving Activated Monomer Mechanism. Cationic Polymerization of Cyclic Ethers Containing Hydroxyl Groups.

M. Bednarek, T. Biedroń, K. Kałużyński, P. Kubisa*, J. Pretula, S. Penczek

Center of Molecular and Macromolecular Studies
Sienkiewicza 112, 90-363 Łódź, Poland

SUMMARY: Cationic polymerization of cyclic ethers containing hydroxyl groups as substituents is discussed in terms of contribution of Active Chain End (ACE) and Activated Monomer (AM) polymerization mechanisms.

Introduction

In the cationic polymerization of cyclic ethers proceeding by Active Chain End (ACE) mechanism, propagation involves nucleophilic attack of oxygen atom of monomer molecule on carbon atom in α- position to the oxygen bearing the positive charge in growing species - tertiary oxonium ions.

$$\text{wwww-CH}_2\text{-O}^\oplus \underset{\text{CH}_2}{\Big)} + \underset{\text{CH}_2}{\text{O} \Big)} \longrightarrow \text{wwww-CH}_2\text{-O} \quad \text{CH}_2\text{-O}^\oplus \underset{\text{CH}_2}{\Big)} \qquad (1)$$

Because linear ether units in polymer are also relatively strong nucleophiles, back-biting and/or transeterification processes involving an attack of oxygen atom of linear unit of the own (cyclization) or foreign chain (so called scrambling) on growing species, cannot generally be avoided.

$$\text{wwww O wwww CH}_2\text{-O}^\oplus \underset{\text{CH}_2}{\Big)} \quad \text{wwww O wwww} \qquad (2)$$

If other nucleophiles are present in the system in sufficient concentration, they may successfully compete with cyclic ether monomer in the reaction with initiating species. This provides a base for Activated Monomer (AM) mechanism of cyclic ethers polymerization [1-3]. AM polymerization is carried out in the presence of hydroxyl group containing compounds and is initiated by protonic acids introduced as such or formed in situ by reaction of e.g. Lewis acid with HO- group,

          CCC 1022-1360/00/$ 17.50+.50/0

therefore, further in the text catalyst will be denoted as "H$^+$".

$$R\text{-}OH + H\text{-}\overset{\oplus}{O} \diagdown CH_2 \longrightarrow R\text{-}O\text{-}CH_2\text{-}O\text{-}H + "H^{\oplus}"$$

$$(3)$$

$$R\text{-}O\text{-}CH_2\text{-}O]_m\text{-}H + H\text{-}\overset{\oplus}{O}\diagdown CH_2 \longrightarrow R\text{-}O\text{-}CH_2\text{-}O]_{n+1}\text{-}H + "H^{\oplus}"$$

In the polymerization proceeding by AM mechanism, due to the absence of active species at the growing chain ends, back-biting is essentially eliminated and linear polymers free of cyclic fraction may be obtained [4,5]. Kinetic studies of AM polymerization of cyclic ethers revealed, that in order to eliminate contribution of ACE mechanism, the instantaneous ratio: [HO-] / [monomer] should be kept above certain level, different for different monomers depending on their nucleophilicity, e.g. above 5 for epichlorohydrin [6]. This may be achieved by carrying out polymerization at monomer starved conditions (i.e. by slow addition of monomer to polymerizing mixture).

This raises an interesting question what is the mechanism of polymerization of cyclic monomers combining both functions i.e. cyclic ether function and hydroxyl function within the same molecule Several monomers of this class are easily available, examples are given below:

$$(4)$$

glycidol

3-hydroxymethyl-oxetane

3-ethyl-3-hydroxy-methyloxetane (EOX)

3,3-bis(hydroxymethyl) oxetane

The presence of HO- group inevitably leads to branching, therefore, in the past polymerization of this group of monomers was not extensively studied [7]. In the recent years, however, an increasing interest in the preparation of highly branched macromolecules and increasing knowledge of AM polymerization mechanism led to renewed interest in the polymerization of this group of cyclic monomers. It may be argued, on the basis of schemes (1) and (2) that the structure of corresponding polymers (degree of branching) should depend on the relative contributions of ACE and AM polymerization mechanisms.

Cationic polymerization of glycidol

Three membered cyclic ether - glycidol is a convenient model for studying the polymerization mechanism because its AM polymerization may lead to isomerized 1-4 units and not only to regular 1-3 units, as in ACE polymerization mechanism [8,9]. 1-3 units contain primary hydroxyl groups while 1-4 units contain secondary hydroxyls. The presence of secondary hydroxyl groups in polymers of glycidol with molecular weights up to 10000 (DP_n = 140) was shown by analysis of ^{29}Si NMR spectra of silylated samples. Two signals at 16.35 and 15.70 ppm δ corresponding to silylated primary and secondary hydroxyl groups respectively were observed in nearly equal intensity, showing that contribution of AM mechanism is significant.

(5)

Dworak et al. estimated the contribution of AM mechanism on the basis of analysis of ^{13}C NMR spectra of polyglycidol, assuming that isomerized 1-4 units and branched units are formed by AM mechanism while linear 1-3 units are formed by ACE mechanism [10]. The results are collected in Table 1.

Table 1. Estimated contribution of AM mechanism in the polymerization of glycidol initiated with different initiators.

Initiator (catalyst)	Contribution of AM, %	DP_n
SnCl$_4$	80	65
BF$_3$ Et$_2$O	60	93
HOSO$_2$CF$_3$, HPF$_6$ Et$_2$O	50	142

Results shown in Table 1 indicate, that there seems to be a correlation between the contribution of AM mechanism and DP_n of the resulting polymers. The values of DP_n correspond to polymers prepared in bulk with concentration of catalyst (protonic acid or Lewis acid) close to 10^{-2} mol/L. The ratio of initial concentrations: $[M]_0$ / $[catalyst]_0$ was therefore above 10^3. The observed polymerization degrees were much lower, and apparently the higher was the contribution of AM mechanism the lower was DP_n.

This raises an interesting question what should be the DP_n values if polymerization proceeded exclusively by AM mechanism. In such a system protonic or Lewis acid would act merely as catalyst (not as initiator) while part of the monomer molecules would act as initiator through their hydroxyl group functions. The degree of polymerization would then be related to relative rates of the competing reactions and not to concentrations of reactants. Assuming, that the terminal oxirane groups can further participate in reaction, the degree of polymerization should in principle increase continuously. This was not observed and the possible explanation is, that oxirane rings may be consumed also in intramolecular reaction with hydroxyl group of the same macromolecule. In such a process the cyclic structure would be formed and the macromolecule would be no longer able to grow after all oxirane rings were consumed. In qualitative terms, such explanation is consistent with the observation, that the higher the contribution of AM mechanism (i.e. mechanism involving the reaction of activated oxirane ring with HO- group) the lower the observed DP_n.

Cationic polymerization of hydroxymethyloxetanes

Vandenberg et al. studied the cationic polymerization of 3-hydroxyoxetane [11]. This monomer gives medium molecular weight polymers (M_n up to 2200) containing both primary (at the ends) and secondary (along the chain) hydroxyl groups and about 5 branching points per chain. From the point of view of the uniformity of structure, more interesting are oxetane derivatives substituted at 3-position with hydroxymethyl groups, because independently on the polymerization mechanism, HO- groups may appear only in $HO-CH_2-C$ units. Until recently, there was essentially no information in the literature on the cationic polymerization of these monomers, except for a brief information in Vandenberg's paper that 3,3-bis(hydroxymethyl)oxetane can be polymerized to low molecular weight oligomers, which were not fully characterized [12].

In the recently published paper, we investigated the cationic polymerization of 3-ethyl-3-hydroxymethyloxetane (EOX) and the structure of polymers in more detail [13]. At the same time, the paper from another group, dealing with the same subject, was published [14].

Below, all possible structures, which may appear in polymers of EOX, are shown:

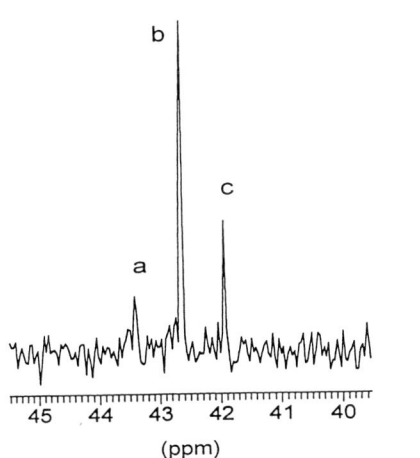

$$(6)$$

Fig. 1 shows expanded region of ^{13}C NMR spectra in which signals of quaternary carbon atoms appear while Fig. 2 shows expanded region of ^{1}H NMR spectra of poly-EOX in which signals corresponding to $-OCH_2O-$ and $-OCH_2OH$ groups can be observed.

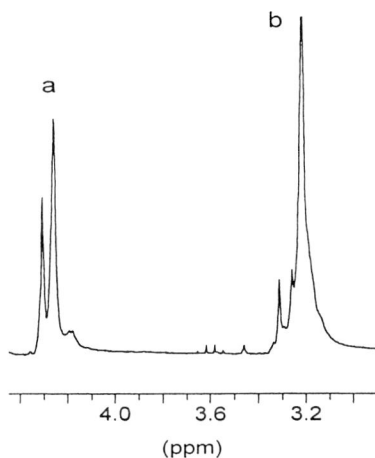

Fig. 1. ^{13}C NMR spectrum of poly-EOX (only the region of quaternary carbon signals shown). Assignments: a:$(-OCH_2)_3CR$, b:$(-CH_2)_2CRCH_2OH$, c: $-OCH_2CR(CH_2OH)_2$.

Fig. 2. ^{1}H NMR spectrum of poly-EOX (only the region of $-CH_2O-$ groups signals shown). Assignments: a: $-OCH_2OH$, b: $-OCH_2O-$

Signals corresponding to three types of units, namely to terminal units containing (except for C$_2$H$_5$- group) two HO-CH$_2$- groups and one -CH$_2$-O-CH$_2$- group bound to quaternary carbon atom, to linear chain units containing one HO-CH$_2$- group and two -CH$_2$-O-CH$_2$- groups bound to quaternary carbon atom and to branched chain units containing three -CH$_2$-O-CH$_2$ groups bound to quaternary carbon atom are clearly observed. In the isolated polymers (after work-up) neither the signals corresponding to terminal unit containing oxetane ring nor the signals corresponding to tertiary oxonium ion involving oxetane ring, could be detected. ^{13}C NMR spectra were recorded in Inverted Gate mode, allowing integration of the spectra. Therefore, the fractions of three observed chain units could be determined from the spectra.

Typical results of the polymerization of EOX are collected in Table 2

Table 2. Cationic polymerization of EOX

Conditions	catalyst, mol%	Yield, precipitated polymer, %	M$_n$	M$_w$ / M$_n$	DP$_n$	(linear + terminal)/ branched units
			gpc in THF			
bulk, -30^{0}C, 48 h	BF$_3$·Et$_2$O 0,2	90	1430	1.70	13	4.0
4 mol/L in CH$_2$Cl$_2$ 25^{0}C, 16 h	BF$_3$·Et$_2$O 0,4	30 (0.5 h) 60 (5.5 h) 90 (48 h)	1140	1.35	10	4.7
bulk, 25^{0}C, 18 h	HOSO$_2$CF$_3$ 0.15	50	1270	1.30	11	high
+10 min. At 200^{0}C	"	100	1890	4,45	18	6.2
+20 min. At 200^{0}C	"	100	insoluble in THF			3.0
+30 min. At 200^{0}C	"	100	Insoluble in typical solvents			

Molecular weights were determined by GPC in THF solution for polymers in which hydroxyl groups were esterified with trifluoroacetic anhydride with standard polystyrene calibration. The results of the determination of molecular weights of branched polyethers, may therefore involve significant errors, they show, however, that independently on reaction conditions and the nature of catalyst, polymers with relatively low molecular weights are formed. Slightly higher (M_n about 3000) but also rather low values were reported by Hult et al. for polymers obtained at 120^0C in bulk with benzyltetramethylsulfonium hexafluoroantimonate as catalyst [14].

As shown by the last two entries of Table 2, heating of the polymers after monomer had been consumed, results in further increase of the degree of branching but at the same time polymers are eventually becoming insoluble in typical organic solvent, which may indicate that cross-linking occurs. To some extent, the increase of the degree of branching may occur through the reaction of terminal oxetane groups, as indicated by the scheme below:

$$-\overset{|}{\underset{|}{C}}-CH_2-OH + \overset{\oplus}{H-O}\diagup\diagup\underset{CH_2-O\sim}{\overset{C_2H_5}{\diagup}} \longrightarrow -\overset{|}{\underset{|}{C}}-CH_2-O\diagup\diagup\underset{O-H}{\overset{C_2H_5\ CH_2-O\sim}{\diagup}} \tag{7}$$

This process cannot, however, lead to cross-linking. The changes observed during heating of the polymer at relatively high temperature in the presence of acid catalysts may rather be related to condensation of the $- CH_2-OH$ groups with elimination of water (it should be mentioned, however, that polymers after neutralization of acid catalysts, are stable at temperature 200^0C).

Mechanism of polymerization of 3-ethyl-3-hydroxymethyloxetane (EOX)

Polymerization of EOX catalyzed by protonic acid (added as such or formed from Lewis acid and HO- group) starts by reaction of protonated EOX:

$$\text{(8)}$$

Reaction 8a, followed by subsequent propagation steps according to ACE mechanism leads to linear chain, although some branching may occur as a result of chain transfer to HO- group of the other macromolecule. Reaction 8b corresponds to AM propagation, but even if reaction proceeds exclusively by this route, branch points may be formed only after certain DP_n is reached, as shown schematically below:

$$(9)$$

Except for a potential branch point introduced at the initiation the possibility of formation of a branch point appear only at the stage of pentamer (reaction of one group denoted as b). Reactions of four groups denoted as l lead to extension of linear chain.

Only after reaching the stage of a pentamer, formation of branch point is possible, and even in this case the probability of forming branch point, assuming equal reactivities of all HO- groups is only 20% (one HO- group out of five). This probability increases with increasing DP_n but decreases with increasing degree of branching (as a result of a formation of branch point on HO-CH$_2$- group in linear unit is consumed and two HO-CH$_2$- end-groups are formed) as shown schematically below:

$$(10)$$

but

This reasoning is consistent with the experimental results presented in Table 2. The degree of branching is rather low and in macromolecules with DP_n between 10 and 13 there is no more than 2-3 branch points. From simple statistical considerations, it follows therefore, that in order to obtain polymers having the structure closer to dendritic structure (i.e. hyperbranched polymers), higher polymerization degrees, than these obtained until now, are needed.

What are the factors limiting the degree of polymerization in the studied systems? For ACE mechanism, chain transfer to HO- group of monomer may lead to termination of particular chain with formation of "proton" which may initiate a new chain. By this reaction, however, oxetane ring would be introduced at the chain end of the terminated macromolecule as shown in scheme below, and such macromonomer should be able to participate in further reaction:

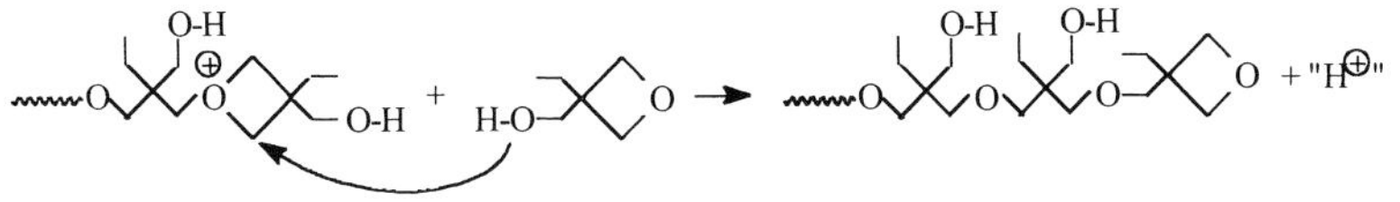

(11)

On the other hand, the same group should be present in all macromolecules growing by AM mechanism and also these macromolecules should undergo further polymerization even after complete consumption of monomer. The possible process by which the growth of macromolecules can be terminated, is the intramolecular reaction of either tertiary oxonium ion with HO- group of its own chain or intramolecular reaction of protonated oxetane end-group with HO- group of its own chain as shown schematically below:

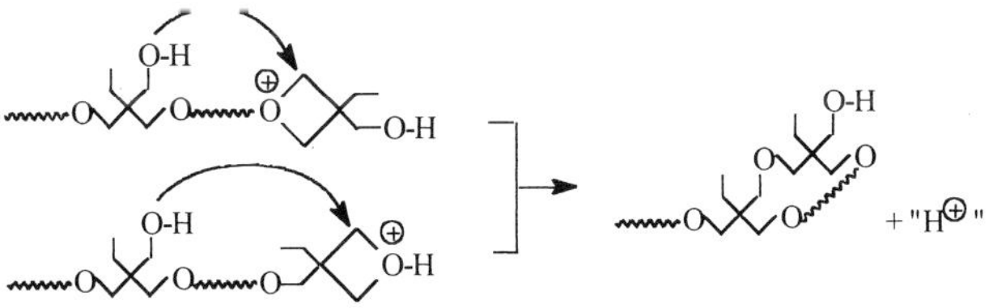

(12)

One of a possible ways to avoid reactions shown in the scheme above, is to carry out the polymerization in the presence of multihydroxyl initiators at the monomer starved conditions. This should lead to increasing contribution of AM mechanism and growing macromolecules would not contain oxetane end-group. Two of the multihydroxyl initiators used in this work are shown in scheme:

$$HO-CH_2 \overset{\overset{\displaystyle CH_2-OH}{|}}{\underset{\underset{\displaystyle C_2H_5}{|}}{C}} CH_2-OH \qquad HO-CH_2 \overset{\overset{\displaystyle CH_2-OH}{|}}{\underset{\underset{\displaystyle CH_2-OH}{|}}{C}} CH_2-O-CH_2 \overset{\overset{\displaystyle CH_2-OH}{|}}{\underset{\underset{\displaystyle CH_2-OH}{|}}{C}} CH_2-OH \qquad (13)$$

As shown by model studies of AM polymerization of oxiranes, high contribution of AM mechanism, even at the monomer starved conditions, could be achieved only to certain limit of DP_n. Above this limit, the increasing participation of reaction of chain units was observed Essentially the same phenomenon was observed for polymerization of EOX with multihydroxy initiators. For [EOX] / [I] ratio equal to 10, the observed M_n = 1100 was still close to the calculated value of M_n = 1170 but for [EOX] / [I] ratio equal to 40 the observed value of M_n = 1970 was already much lower than calculated value of M_n = 4650. The degree of branching was not significantly higher than in the polymerization without added multihydroxyl initiator although increase of the degree of branching with increasing DP_n was clearly observed.

Table 3. The dependence of the degree of branching on $[EOX]_{consumed}$ / $[Trihydroxyl\ initiator]_0$

$[EOX]_{consumed}$ / $[Trihydroxyl\ initiator]_0$	4	6	8	10
Terminal + linear units / branched units	7.1	5.4	5.0	3.8

Conclusions

Cationic polymerization of 3-ethyl-3-hydroxymethyloxetane leads to medium molecular weight, branched polyethers containing $-CH_2-OH$ groups both at the chain ends and along the chain. The number of branch points is related to the degree of polymerization; up to now polymers with M_n up to 2000 were obtained and characterized. The extension of the available molecular weight range requires better understanding of polymerization kinetics, in particular the extend of participation of ACE and AM mechanism in the process.

References

1. K. Brzezińska, R. Szymański, P. Kubisa, S. Penczek, Makromol. Chem., Rapid Commun. 7, 1 (1986)
2. S.Penczek, P.Kubisa, "Cationic Ring-Opening Polymerization" in "Ring-Opening Polymerization, mechanism, Catalysis and Utility" J.D.Brunelle, Ed., Hanser-Verlag, New York, 1993, p.13-86
3. S.Penczek, H.Sekiguchi, P.Kubisa, "Activated Monomer Polymerization of Cyclic Monomers" in "Frontiers of Polymer Science", K.Hatada, O.Vogl, Eds., 1996, p.199-222.
4. M. Wojtania, P. Kubisa, S. Penczek, Makromol. Chem., Macromol. Symp. 6, 201 (1986)
5. T. Biedroń, P. Kubisa, S. Penczek, J. Polym. Sci., Polym. Chem. Ed., 29, 619 (1991)
6. T. Biedroń, R. Szymański, P. Kubisa, S. Penczek, Makromol. Chem., Macromol. Symp., 32, 155 (1990)
7. J. A. Wojtowicz, R. J. Polak, J. Org. Chem. 38, 2061 (1973)
8. E. J. Vandenberg, J. Polym. Sci., Polym. Chem. Ed., 23, 915 (1985)
9. R. Tokar, P. Kubisa, S. Penczek, A. Dworak, Macromolecules, 27, 320 (1994)
10 A. Dworak, W. Wałach, B. Trzebicka, Macromol. Chem. Phys. 196, 1963 (1995)
11. E. J. Vandenberg, J. C. Mullis, R. S. Juvet Jr., T. Miller, R. A. Nieman, J. Polym. Sci., Polym. Chem. Ed., 27, 3113 (1989)
12. E. J. Vandenberg, J. C. Mullis, R. S. Juvet Jr., J Polym. Sci., Polym. Chem. Ed., 27, 3083 (1989)
13. M. Bednarek, T, Biedroń, J. Heliński, K. Kałużyński, P. Kubisa, S. Penczek, Macromol. Rapid Commun. 20, 369 (1999)
14. H. Magnusson, E. Malmström, A. Hult, Macromol. Rapid Commun. 20, 453 (1999)

SYNTHESIS OF A NOVEL POLYMERIC CARBOHYDRATE VIA REGIO- AND STEREOSELECTIVE CYCLOPOLYMERIZATION OF 1,2:5,6-DIANHYDROHEXITOL

Kazuaki Yokota,*[1] Toyoji Kakuchi,[2] Toshifumi Satoh, [1] Stoshi Umeda, [‡][1] and Masatoshi Kamada [1]

[1] Division of Molecular Chemistry, Graduate School of Engineering, Hokkaido University, Sapporo 060-8628, Japan

[2] Division of Bioscience, Graduate School of Environmental Earth Science, Hokkaido University, Sapporo 060-0810, Japan

Abstract: The selective cyclopolymerization of 1,2:5,6-dianhydrohexitols corresponding to diepoxides was a new synthetic strategy for polycarbohydrates, though the polymer is a lack of the anomeric linkage which is found in the naturally occurring polysaccharides. 1,2:5,6-Dianhydro-3,4-di-*O*-methyl-D-mannitol, L-iditol, and D-glucitol were polymerized using *t*-BuOK and BF$_3$•OEt$_2$ to produce the polymers consisting of five-membered rings. On the other hand, the polymers consisting of six-membered rings were obtained by the cationic and anionic polymerizations of meso allitol and galactitol monomers, respectively.

INTRODUCTION

Methods for synthesizing polycarbohydrates fall into three categories: sequential condensation polymerization, ring-opening polymerization, and enzymatic polymerization. The cyclopolymerization of diepoxides derived from hexitols, namely, 1,2:5,6-dianhydrohexitols presents a new preparative method for polycarbohydrates [Refs. 1,2]. Diepoxides with suitable structure come in a type of monomers to undergo cyclopolymerization. 1,2:5,6-Dianhydrohexitols derived from naturally occurring hexitols correspond to the diepoxides. There are ten hexitols that contain four pairs of optical enantiomers and two meso forms. Two pairs of D- and L-isomers in mannitol and iditol are C$_2$ symmetric and the other pairs are asymmetric. D-Glucitol with C$_1$ symmetry consists of

---

*To whom correspondence should be addressed.
‡ Research Fellow of the Japan Society for the Promotion of Science.

© WILEY-VCH Verlag GmbH, D-69469 Weinheim, 2000   CCC 1022-1360/00/$ 17.50+.50/0

the structure combined two halves of D-mannitol and L-iditol. D-Altritol is constructed by two halves of meso allitol and galactitol. 1,2:5,6-Dianhydro-3,4-di-O-methyl-D-mannitol and L-iditol (**1** and **3**) of which the two epoxy groups in a molecule are equal in reactivity polymerize to yield the polycarbohrates without the anomeric linkage, unlike the naturally occurring polysaccharides. 1,2:5,6-Dianhydro-3,4-di-O-methyl-D-glucitol (**7**) possesses two epoxy groups whose reactivities are nonequivalent, thus yielding two directions for polymerization. 1,2:5,6-Dianhydro-3,4-di-O-methyl-allitol and galactitol (**11** and **13**) with a plane of symmetry through the C3-C4 bond form the polymers consisting of a pair of enantiomeric cyclic units.

Cyclopolymerization of 1,2:5,6-dianhydro-3,4-di-O-methyl-D-mannitol (1)

The anionic cyclopolymerization of **1** was carried out in THF and toluene at 60 °C for 48 h using *t*-BuOK to yield the polymer (**2a**) [Refs. 3,4]. The reaction system was homogeneous up to complete consumption of the monomers. For the polymerization with a [**1**]/[*t*-BuOK] ratio of 40 in toluene, the $\overline{M}_n$ of the polymer attained to 12900 ($\overline{P}_n = 74$). The sparing solubility of *t*-BuOK in toluene lowered the initiator efficiency. The complexation of *t*-BuOK with 18-crown-6 increased the initiator efficiency from 0.3 to 1.0. Since the linearity in the plot of $\overline{P}_n$ versus conversion showed little participation of side reactions in polymerization, e.g., chain transfer to monomer, the living-like nature of the system was found [Refs. 4,5]. Polymer **2a** was sticky semisolid and soluble in common organic solvents.

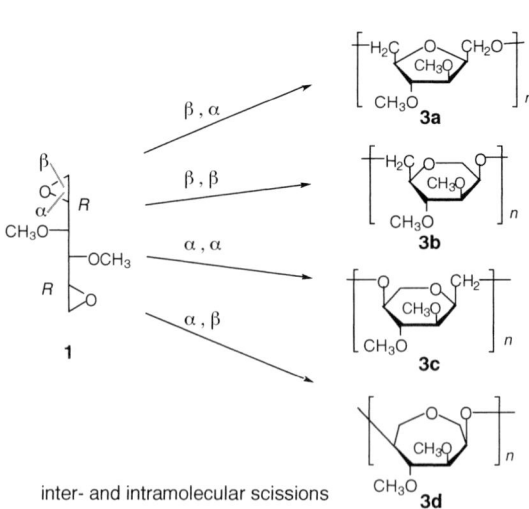

The specific rotation $([\alpha]_{546}^{22})$ of the polymer ranged from +72.2 to +93.9 deg·cm^2·dag^{-1} (concentration, 10 g·dm^{-3} in CHCl$_3$).

There are four possible cyclic units (**3a** - **3d**) by combination of the inter- and intramolecular reactions in the cyclopolymerization of **1** (Scheme 1). To confirm the structure of polymer **2a**, therefore, model compound **4**,

Scheme 1

2,5-anhydro-1,3,4,6-tetra-O-methyl-D-glucitol was synthesized by hydrolysis of **1** and then treatment with dimethyl sulfate. The signals for the ^{13}C NMR spectrum of polymer **2a** fairly agreed with the chemical shifts for the model compound. Polymer **2a**, therefore, was exclusively $(1{\to}6)$-2,5-anhydro-3,4-di-O-methyl-D-glucitol (**3a**).

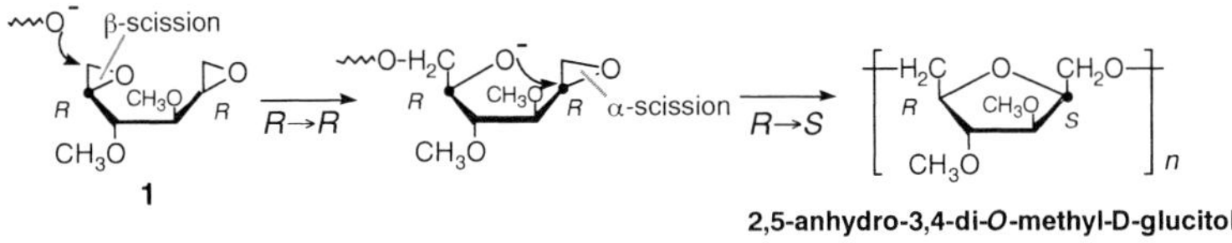

Scheme 2

For the cyclopolymerization of **1** using t-BuOK, the stereochemically-controlled polymer **2a** should be produced through the mechanism proposed in Scheme 2. For the intermolecular reaction, the growing alkoxy anion cleaves the β-bond of the first epoxide, resulting in retention of the R configuration of the α-carbon. On the other hand, for the intramolecular cyclization, the alkoxy anion cleaves the α-bond of the second epoxide to form a 5-membered ring. For the stereoselectivity during the cyclization, the configuration of the α-carbon is inverted from R to S. The cyclopolymerization **1** using t-BuOK was highly regio- and stereospecific through β,α-scissions in the inter- and intramolecular reactions.

With $BF_3{\cdot}OEt_2$ in toluene or methylene chloride, the n-hexane-insoluble polymer (**2b**) was obtained together with n-hexane-soluble low molecular weight products [Ref. 6]. Polymer **2b** was a sticky semisolid soluble in $CHCl_3$, MeOH, THF, and H_2O and was optically active with a specific rotation $[\alpha]_{546}^{22}$ in the range from +41.3 to +71.3 deg·cm^2·dag^{-1}. The $\overline{M}_n$ was 1000 - 3400 ($\overline{P}_n$ = 6 - 20). Although polymer **2b** contained some irregular cyclic units as a minor component together with 2,5-anhydro-D-glucitol unit **3a**, the polymerization also proceeded mainly through β,α-scissions.

Cyclopolymerization of 1,2:5,6-dianhydro-3,4-di-O-methyl-L-iditol (5)

The polymerization systems of **5** using t-BuOK were homogeneous up to a very high conversion [Ref. 7]. The polymer (**6a**) had the $\overline{M}_n$ ranging from 2600 to 6100 ($\overline{P}_n$ = 15 - 35). Chain transfer to monomer placed the upper limit to the $\overline{M}_n$ of polymer **6a** [Ref. 8]. In the ^{13}C NMR spectrum of polymer **6a**, the signals agreed fairly well with the chemical shifts for model compound **4**. The cyclic constitutional unit in polymer **6a**,

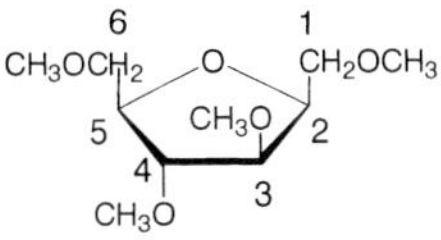

4

thereby, was recognized as 2,5-anhydro-3,4-di-O-methyl-D-glucitol, analogous to polymer **2a**. The formation of the 2,5-anhydro-3,4-di-O-methyl-D-glucitol units is based on β,α-scissions of the two epoxy groups in the monomer. The anionic cyclopolymerization of **5**, thus, is regio- and stereoselective, like that of **1**. However, there is an essential difference in structure between polymers **2a** and **6a**. Although both polymers have apparently the same constitutional repeating units, their units differ from one another in direction. The ^{13}C NMR spectra of polymers **2a** and **6a** showed the small signals attributable to the ends in the polymer chain. Polymer **2a**, therefore, is (1→6)-2,5-anhydro-3,4-di-O-methyl-D-glucitol with tert-butoxy and hydroxymethyl groups at both ends, but polymer **4a** by the (6→1)-bonded glucitol, as shown in Scheme 3. The copolymerization between monomers **1** and **5** offered a solution to the problem of direction.

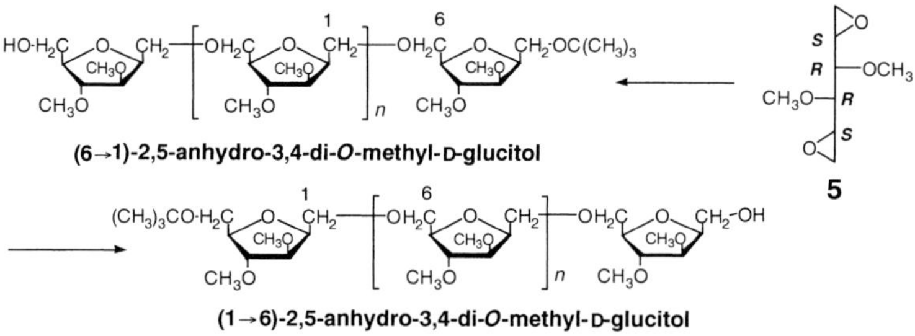

(6→1)-2,5-anhydro-3,4-di-O-methyl-D-glucitol

5

1

(1→6)-2,5-anhydro-3,4-di-O-methyl-D-glucitol

Scheme 3

In the cationic polymerization of monomer **5** with $BF_3 \cdot OEt_2$, the yields and $\overline{M}_n$s for the polymer (**6b**) were lower than those for **2b** [Ref. 6]. The ^{13}C NMR spectrum of polymer **6b** indicated that the major structure of the polymer was identical to polymer **6a**, but some irregular units were formed.

Cyclopolymerizations of 1,2:5,6-dianhydro-3,4-di-O-methyl-D-glucitol (7)

The anionic polymerization of **7** using *t*-BuOK yielded the polymer (**8a**) with the $\overline{M}_n$ in the range of 4530 to 5510 ($\overline{P}_n$ = 26 - 32) [Ref. 9]. The polymer (**8b**) obtained by the cationic polymerization with $BF_3 \cdot OEt_2$ had the $\overline{M}_n$s of 3770 ($\overline{P}_n$ = 22). The specific rotations ($[\alpha]_{546}^{22}$) were +69.2 to +81.4 deg·cm²·dag⁻¹ for polymer **8a**, and +40.6 to +62.7 deg·cm²·dag⁻¹ for polymer **8b**. The characteristic signals due to the epoxy groups in the ¹H NMR spectra of both polymers completely disappeared. The ¹³C NMR spectrum of polymer **8a** comprised of the signals similar to those of the model compounds, 2,5-anhydro-3,4-di-O-methyl-Dt-BuOK

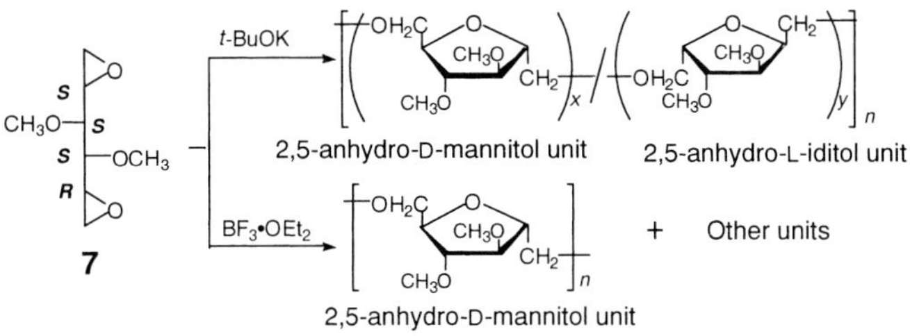

OH$_2$C CH$_3$O CH$_2$ CH$_3$O
2,5-anhydro-D-mannitol unit

O CH$_2$ CH$_3$O OH$_2$C CH$_3$O
2,5-anhydro-L-iditol unit

n

x y

7

BF$_3$•OEt$_2$

OH$_2$C CH$_3$O CH$_2$ CH$_3$O
2,5-anhydro-D-mannitol unit n

+ Other units

Scheme 4

mannitol (**9**) and L-iditol (**10**). Polymer **6a**, thus, consists of two kinds of cyclic repeating units, as shown in Scheme 5. The ratio of the D-mannitol unit to the L-iditol unit (x/y) was 0.40/0.60 for the polymerization in toluene. The ratio changed from 0.27/0.73 in the system of 18-crown-6/toluene to 0.47/0.53 in DMSO according to the polymerization conditions. On the other hand, in the structure of polymer **8b**, there might exist six- and seven-membered rings as minor constitutional units together with the 2,5-anhydro-3,4-di-O-methyl-D-mannitol as major unit.

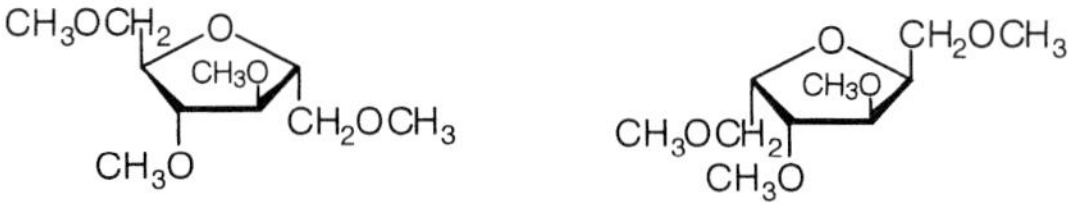

CH$_3$OCH$_2$ CH$_3$O CH$_2$OCH$_3$ CH$_3$O

9

O CH$_2$OCH$_3$ CH$_3$O CH$_3$OCH$_2$ CH$_3$O

10

The anionic cyclopolymerization of **7** produced the polymer consisting of two kinds of cyclic repeating units. The β_{12},α_{56}-scissions form the cyclic D-mannitol unit with S,S-configuration, and the β_{56},α_{12}-scissions form the cyclic L-iditol unit with R,R-configuration (Scheme 5). As a result, the two processes competitively proceeded to produce the

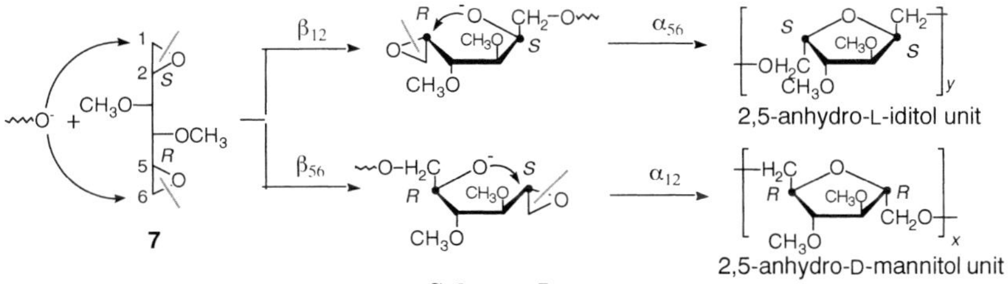

Scheme 5

copolymer consisting of 2,5-anhydro-3,4-di-O-methyl-D-mannitol and L-iditol units. Although involving two processes, the polymerization was regio- and stereoselective in each process. On the other hand, the cationic cyclopolymerization of **7** induced the intermolecular reaction preferentially at the 1,2-epoxide moiety as a nucleophile. Therefore, the formation of 2,5-anhydro-3,4-di-O-methyl-D-mannitol results from β_{56},α_{12}-scissions.

Cyclopolymerizations of meso 1,2:5,6-dianhydro-3,4-di-O-methyl-allitol (11) and galactitol (13)

The anionic polymerization systems using t-BuOK were homogeneous for **11** and heterogeneous for **13** [Ref. 10]. The obtained polymers (**12a** and **14a**) were soluble in chloroform and THF. For **12a** obtained in toluene, the $\overline{P}_n$s were 7.6, 14.4, and 25.1 relative to the values of 5, 10, and 20 calculated from the ratio of [M]/[t-BuOK]. The polymerization of **11**, thus, had a living-like nature similar to that of **1**. On the other hand, the polymerization of **13** suggested some participation of the chain transfer to monomer similar to those of **5** and **7**. The cationic polymerizations of **11** and **13** with BF$_3\cdot$OEt$_2$ gave the polymers (**12b** and **14b**) with the lower $\overline{P}_n$s of 13 and 6, respectively. The extent of cyclization (f_c) was 1.0 for each of polymers **12a**, **14a**, and **11b**, but 0.89-0.93 for **14b**. All of the polymers consist of a pair of enantiomeric cyclic units, thus being optically inactive.

Hydrolysis of **11** and **13** yielded 2,5-anhydro-1,3,4,6-tetra-O-methyl-DL-altritol (**15**), 1,5-anhydro-2,3,4,6-tetra-O-methyl-DL-allitol (**16**), and 1,5-anhydro-2,3,4,6-tetra-O-methyl-DL-galactitol (**17**) corresponding to model compounds for the constitutional units. The signals

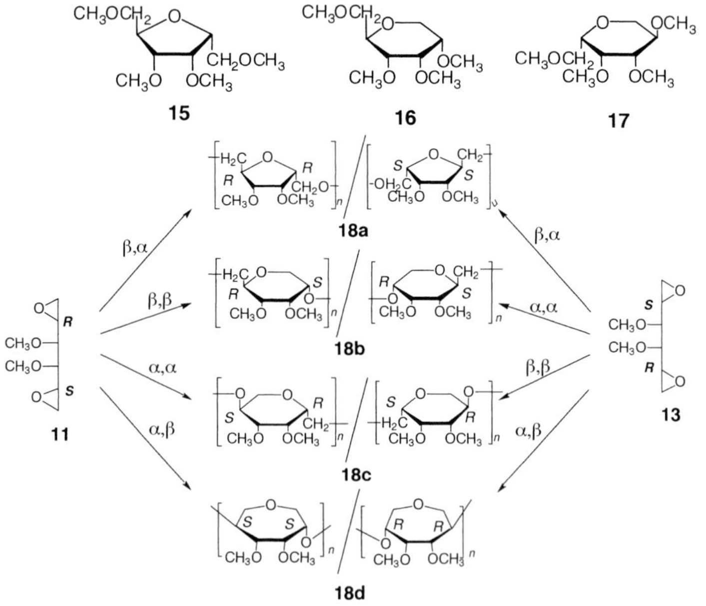

Scheme 6

observed in the ^{13}C NMR spectra of the polymers were split into two or more bands, which fact should be caused by the diversity of sequences of the two enantiomeric units. The signals for polymers **12a** and **14a** agreed very closely with those for **15** and **17**, respectively. The structure of the polymers was **18a** of five-membered rings for polymer **12a**, in contrast to **18c** of six-membered rings for polymer **14a** (Scheme 6).

The ^{13}C NMR spectra of polymers **12b** and **14b** were somewhat complex, which suggested that their structures were composed of plural repeating units. For polymer **12b**, the signals could be divided into two groups compatible with the spectra of **15** and **17**. Polymer **12b**, thus, had a structure consisting of **18c** as major unit and **18a** as minor unit. Polymer **14b** comprised **18a** and **18b** of five- and six-membered rings, respectively. In addition, the other unknown units along with the uncyclized unit were contained in the polymer.

During the anionic cyclopolymerizations of **11**, the constitutional unit **18a** was formed through β,α-scissions. The equality of reactivity at the 1,2- and 5,6-epoxides yielded a pair of enantiomeric units in polymer **12a**. On the other, the anionic polymerization of **13** proceeded mainly through β,β-scissions to form **18c**. The growing anion in cyclization should avoid the unfavorable conformation due to the eclipsed arrangements of three neighboring substituents in the five-membered ring to convert its conformation into the favorably staggered arrangements in the six-membered ring (Scheme 7).

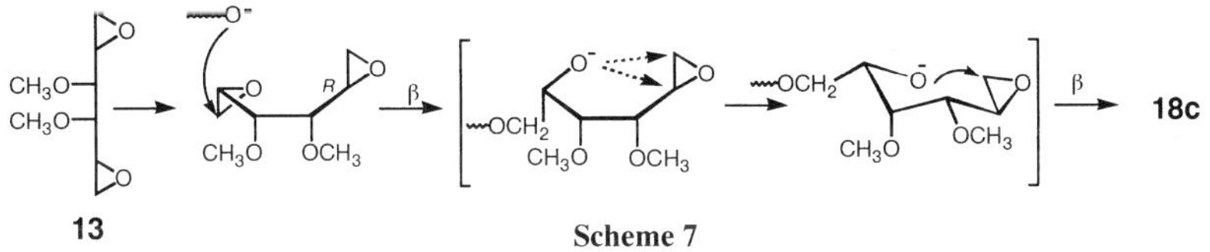

Scheme 7

The cationic cyclopolymerization of **11** comprised the process through α,α-scissions to form **18c**. The steric crowd in the growing oxonium ion is responsible for the formation of six-membered rings. The α,α-scissions in the cationic polymerization of **11** constructed the same constitutional units of six-membered rings as the β,β-scissions in the anionic polymerization of **13**.

Characteristics of (1→6)-2,5-anhydro-3,4-di-*O*-methyl-D-glucitol as a new macromolecular ionophore

The naturally occurring ionophore includes the polyether antibiotics such as monensin and nigericine consisting of a formally linear array of tetrahydrofuranyl and tetrahydropyranyl rings. The regio- and stereoselective cyclopolymerization of monomer **1** yielded the polycarbohydrate consisting of tetrahydrofuranyl rings.

Polymer **2** strongly complexed with such larger organic cations as methylene blue and rhodamine 6G besides alkali metal ions, unlike the crown ether[Ref. 2].   In the transport experiment using the HPF_6 salt of methyl α-amino esters $(RCH(CO_2CH_3)NH_3^+ \cdot PF_6^-)$, the steric requirements for higher enantioselectivity were the longer alkyl substituents at the 3,4 positions for the polymer and the bulkiness of the chiral center for the guest [Refs. 11,12]. Having a living-like character, the anionic polymerization of monomer **1** could be used for preparing stationary phase bound on silica gel for chromatographic optical resolution of racemic amines and amino acids [Ref. 13]

CONCLUSIONS

The anionic polymerizations of 1,2:5,6-dianhydrohexitols were highly regio- and stereoselective to produce the polycarbohydrates consisting of five-membered rings for D-mannitol, L-iditol, D-glucitol, and meso allitol monomers and six-membered rings for meso galactitol.   The polymer consisting of six-membered rings was formed also by the cationic polymerization of allitol monomer.   The same structures of six-membered rings were obtained through β,β-scissions for galactitol monomer and α,α-scissions for allitol monomer.

REFERENCES

(1)   K. Yokota, O. Haba, T. Satoh, T. Kakuchi, *Macromol. Chem. Phys.*, **196**, 2383 (1995).
(2)   H. Hashimoto, T. Kakuchi, K. Yokota, *J. Org. Chem.*, **56**, 6470 (1991).
(3)   T. Satoh. K. Yokota, T. Kakuchi, *Macromolecules*, **28**, 4762 (1995).
(4)   T. Satoh, T. Hatakeyama, S. Umeda, H. Hashimoto, K. Yokota, T. Kakuchi, *Macromolecules*, **29**, 3447 (1996).
(5)   T. Hatakeyama, M. Kamada, T. Satoh, K. Yokota, T. Kakuchi, *Macromolecules*, **31**, 2889 (1998).
(6)   .T. Kakuchi, T. Satoh, S. Umeda, H. Hashimoto, K. Yokota, *Macromolecules*, **28**, 5643 (1995).
(7)   T. Satoh, T. Hatakeyama, S. Umeda, M. Kamada, K. Yokota, T. Kakuchi, *Macromolecules*, **29**, 6681 (1996).
(8)   T. Hatakeyama, M. Kamada, T. Satoh, K. Yokota, T. Kakuchi, *Kobunshi Ronbunshu*, **50**, 710 (1997).
(9)   T. Satoh, T. Hatakeyama, S. Umeda, K. Yokota, T. Kakuchi, *Polym. J.*, **28**, 520 (1996).
(10) M. Kamada, T. Satoh, K. Yokota, T. Kakuchi, *Macromolecules*, in press.
(11) T. Kakuchi, Y. Harada, T. Satoh, K. Yokota, H. Hashimoto, *Polym.*, **35**, 204 (1995).
(12) T. Kakuchi, T. Satoh, S. Umeda, J. Mata, K. Yokota, *CHIRALITY*, **7**, 136 (1995).
(13) T. Kakuchi, T. Satoh, H. Kanai, S. Umeda, T. Hatakeyama, K. Yokota, *Enantiomer*, **2**, 273 (1997)

Living Ring-Opening Polymerization
Based on Neighboring Group Participation

Takeshi Endo,* Fumio Sanda, Wonmun Choi

Research Laboratory of Resources Utilization, Tokyo Institute of Technology

4259 Nagatsuta-cho, Midori-ku, Yokohama 226-8503, Japan

SUMMARY: Cationic ring-opening polymerization of a five-membered cyclic dithiocarbonate having benzoxymethyl group; 5-benzoxymethyl-1,3-oxathiolane-2-thione, was carried out with TfOH or TfOMe as an initiator in PhCl at rt – 60 °C. The molecular weight distribution (M_w /M_n) of the polymer obtained with TfOMe was very narrow even at 60 °C (M_w /M_n 1.14), and the Mn value of the polymers estimated by GPC was in good agreement with the molecular weight determined from ^{1}H-NMR. The living nature of the polymerization was confirmed by the conversion dependence of the Mn (M_w /M_n) and the correlation of the experimental and theoretical M_n (M_w /M_n) values.

Introduction

Since the discovery of anionic living polymerization,[1] the chemistry of living polymerization has been developed in the field of coordination,[2] metathesis,[3] radical,[4] and cationic polymerizations in addition to anionic polymerization. Considerable advances have been achieved in living cationic polymerization for a variety of vinyl monomers such as vinyl ethers,[5] isobutene,[6] styrene,[7] and *N*-vinylcarbazole.[8] Living cationic polymerization is based on the stabilization of a growing carbocation by an added base or a counter anion. Meanwhile, some cyclic monomers such as tetrahydrofuran[9] and oxazoline[10] can undergo cationic living ring-opening polymerization with a stabilized propagating polymer end, because the chain transfer reaction of these cyclic monomers is unfavorable. Recently, we have reported the first example of selective cationic isomerization and ring-opening polymerization of five-membered cyclic dithiocarbonates (**1**). The monomer **1** selectively isomerizes to **4** in the presence of Lewis acids such as ZnCl$_2$ and SnCl$_4$, and protonic acids such as CF$_3$SO$_3$H (TfOH) and CH$_3$SO$_3$H as the catalysts, whereas **1** selectively polymerizes with CF$_3$SO$_3$Me (TfOMe) and CF$_3$SO$_3$Et (TfOEt) as the initiators to afford the corresponding polydithiocarbonates (**5**) (Scheme 1).[11] The formation of a cyclic oxonium cation (**2**) and a cyclic carbenium cation (**3**) has been confirmed in the reactions of **1** with TfOH and TfOMe, respectively (Scheme 1).[12]   The selectivity of the cationic isomerization and polymerization of **1** is attributable to the different intermediates depending on the catalysts.

Scheme 1

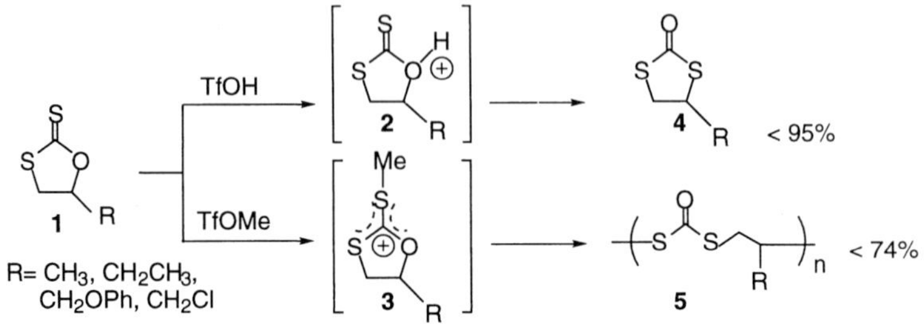

R= CH₃, CH₂CH₃, CH₂OPh, CH₂Cl

Neighboring group participation plays an important role in selective chemical synthesis of oligosaccharides[13] and regiochemical control on the ring-opening of oxirane by nucleophiles.[14] If this neighboring group participation is employed to stabilize a propagating polymer end, a new class of living polymerization will be constructed.   This work deals the first example of a controlled living cationic ring-opening polymerization of a five-membered cyclic dithiocarbonate having a benzoxymethyl group (**1a**) based on the stabilization of the growing carbocation by neighboring group participation.

Results and Discussion

The five-membered cyclic dithiocarbonate (**1a**) was synthesized by the reaction of the corresponding oxirane and CS_2 in the presence of LiBr catalyst according to the previously reported method.[15]   The cationic polymerization of **1a** was carried out under various conditions to give the corresponding polymers as summarized in Scheme 2 and Table 1. TfOH as well as TfOMe selectively gave the polymer, which was completely different from the other cyclic dithiocarbonates (**1**).   It is noteworthy that the molecular weight distributions (M_w/M_n) of the polymers obtained with TfOMe are very narrow even at 60 °C (M_w/M_n 1.14). The M_n values of the polymers estimated by GPC based on polystyrene calibration were in good agreement with the molecular weights determined from the ^{1}H-NMR peak integration ratio of the S-Me group at the initiating end.   After the complete consumption of **1a**, the polymerization took place again when the same amount of **1a** was introduced in the reaction mixture.   The M_n of the polymer increased in direct proportion to the monomer conversion and showed a good agreement with the molecular weight calculated by NMR (Fig. 1).

Scheme 2

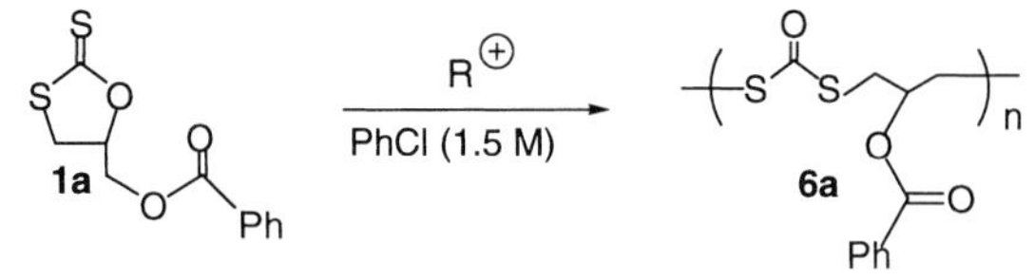

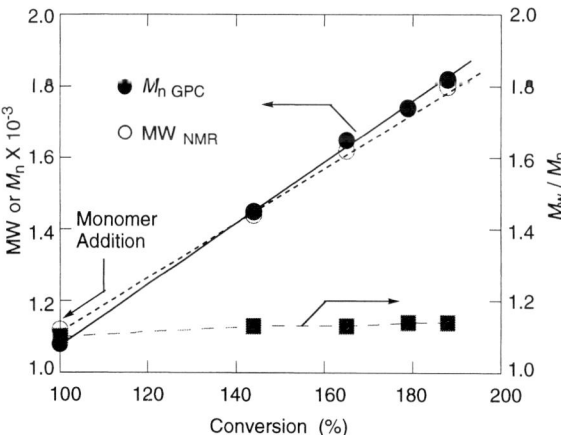

Table 1. Cationic Polymerization of **1a**.

run	init (mol %)	temp (°C)	time (min)	conv [a] (%)	yield (%)	MW $_{NMR}$ [a]	M_n GPC [b]	Mw/Mn [b]
1	TfOH (2)	rt	240	60	55 [c]	-	24600	1.31
2	TfOH (2)	60	60	100	98 [d]	-	16700	1.22
3	TfOMe (2)	rt	480	93	94 [d]	12900	13200	1.10
4	TfOMe (2)	45	30	73	73 [d]	9100	9300	1.09
5	TfOMe (2)	45	90	100	100 [d]	12300	12700	1.10
6	TfOMe (3)	60	60	100	100 [d]	8500	9900	1.14

a) Estimated by ^{1}H-NMR.   b) Estimated by GPC eluted by THF based on polystyrene standards.   c) Isolated by preparative HPLC.   d) *n*-Hexane-insoluble part.

Fig. 1: Conversion dependence of the M_n and M_w /M_n of **6a** obtained by the polymerization of **1a** with TfOMe (2.5 mol %) in PhCl (1.5 M) at 30 °C.

The polymerization of **1a** was carried out with various amounts of TfOMe (0.8 - 8 mol %) at room temperature to confirm the living nature of the polymerization. The M_n of the polymer agreed well with the theoretical value, although the molecular weight distribution was slightly broad with 0.8 mol % of TfOMe (M_n 32600, M_w /M_n 1.18), as shown Fig. 2. Further, the polymerization was quenched with myristyltrimethylammonium bromide to examine the chain-end functionalization of the polymer. The obtained polymer showed ^{1}H-NMR signals assignable to initiating S-Me and terminating bromomethyl end group protons, where the functionality of the terminating end group was 92%, supporting the living nature of the polymerization.

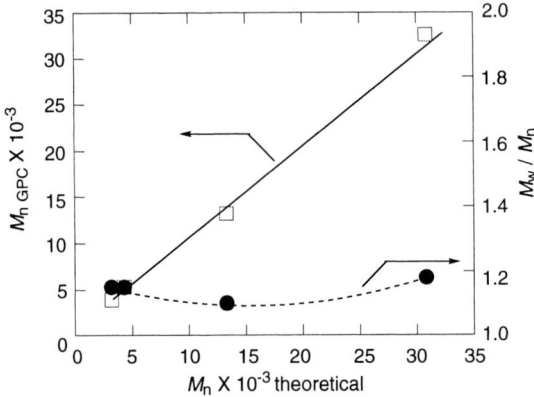

Fig. 2: Correlation of the theoretical and experimental M_n and M_w/M_n of **6a** obtained by the polymerization of **1a** with TfOMe (0.8 - 8 mol %) in PhCl (1.5 M) at room temperature.

The structure of the polymer was confirmed by IR, ^{1}H-NMR, and ^{13}C-NMR spectroscopy besides elemental analysis. Fig. 3 shows the ^{1}H-NMR spectrum of the polymer obtained by the polymerization of **1a** with TfOMe (3 mol %) at 60 °C for 1 h (run 6 in Table 1). In the ^{1}H-NMR spectrum of the polymer, the signal at 4.7 ppm of the α-methylene protons of benzoxy group completely disappeared, and signal *b* assignable to α-methine proton of the benzoxy group appeared at 5.2 ppm. No signal was observed at 4.5-5 ppm, which was expected for the α-methylene protons of the benzoxy group in **5a**. Consequently, it can be concluded that the structure of the polymer is not **5a** but **6a**, which is supported by IR and ^{13}C-NMR spectroscopy. It is quite surprising that the polymer structure is different depending on the substituent on the monomer.

[A]

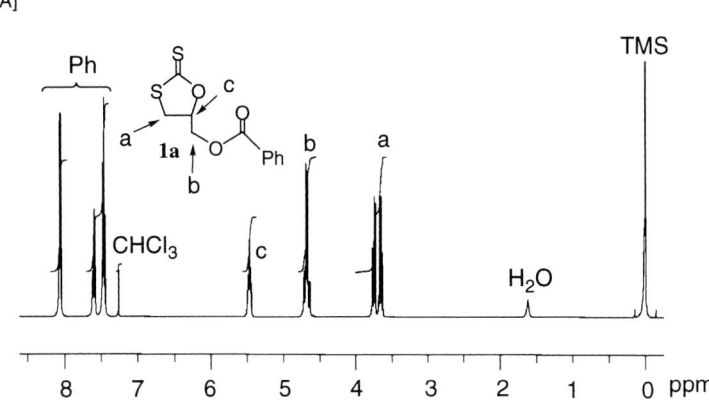

[B]

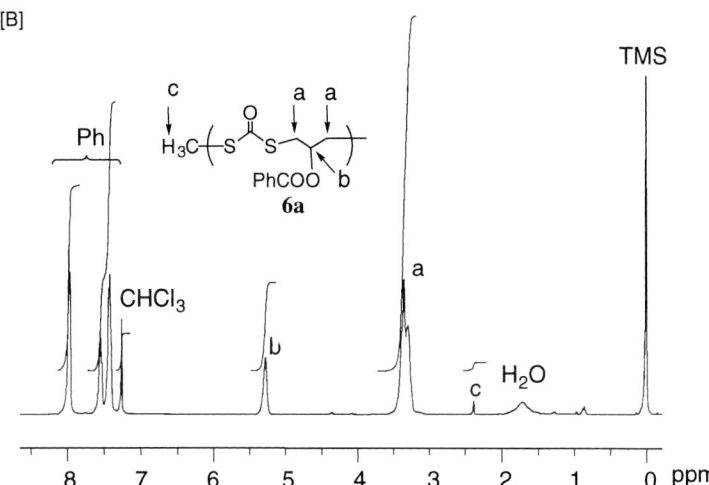

Fig. 3: ¹H-NMR (400 MHz) spectra of **1a** [A] and **6a** [B] obtained by the polymerization of
1a with TfOMe (3 mol %) in PhCl (1.5 M) at 60 °C for 1 h.

The ¹H- and ¹³C-NMR spectra were measured for the mixture of **1a** with TfOMe (1.2 eq) in
$CDCl_3$ at room temperature to examine the possibility of neighboring group participation in
the polymerization. The formation of a carbenium cation (**3a'**, 88%, calculated from ¹H-
NMR) was confirmed with a small amount of a carbenium cation (**3a**, 12%) as shown in Fig.
4.

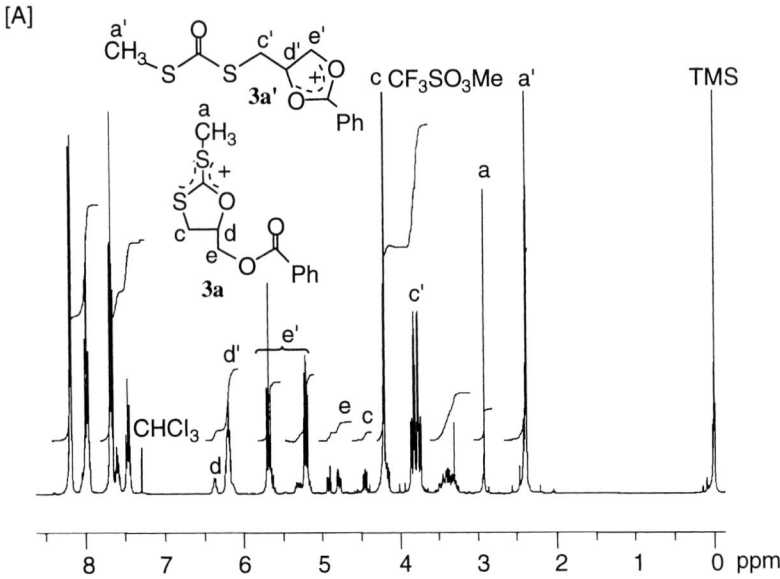

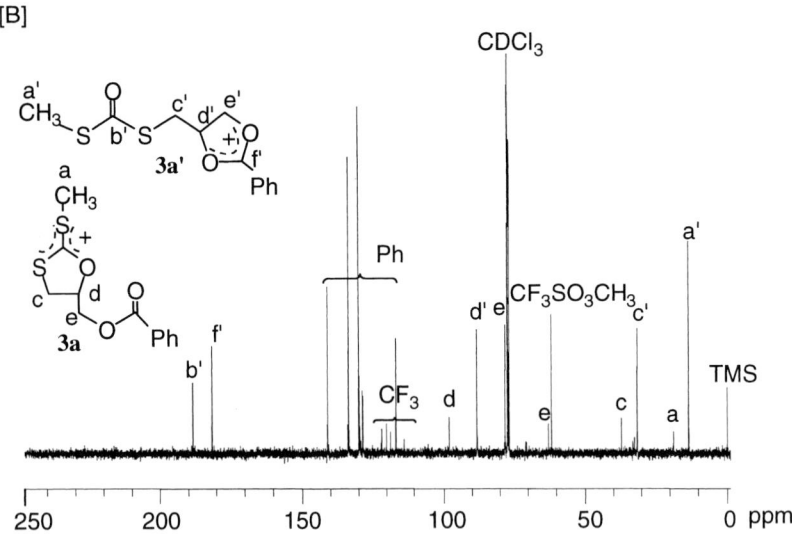

Fig. 4: [A] ^{1}H-NMR (400 MHz) and [B] ^{13}C-NMR (100 MHz) spectra of the mixture of **1a** and TfOMe (1.2 eq) in CDCl$_3$ at room temperature.

Scheme 3 illustrates a plausible mechanism of the polymerization of **1a**. The monomer **1a** forms an oxonium cation (**2a**) and a carbenium cation (**3a**) by protonation or methylation, followed by isomerization to yield a more stable carbenium cation (**3a'**) stabilized by two oxygen atoms and phenyl group. The results suggest that path C selectively proceeds from **3a'** to afford the polymer (**6a**) among the three possible paths; path A to afford the isomer (**4a**), path B to afford the polymer (**5a**) and path C. Path C may be more favorable than paths A and B probably due to steric factors. The stability of **3a'** plays an important role on the selectivity of the polymerization and isomerization, namely, the intramolecular isomerization from **2a** may be suppressed by the formation of the more stable benzoxonium cation (**3a'**) than **2a** and **3a**.

Scheme 3

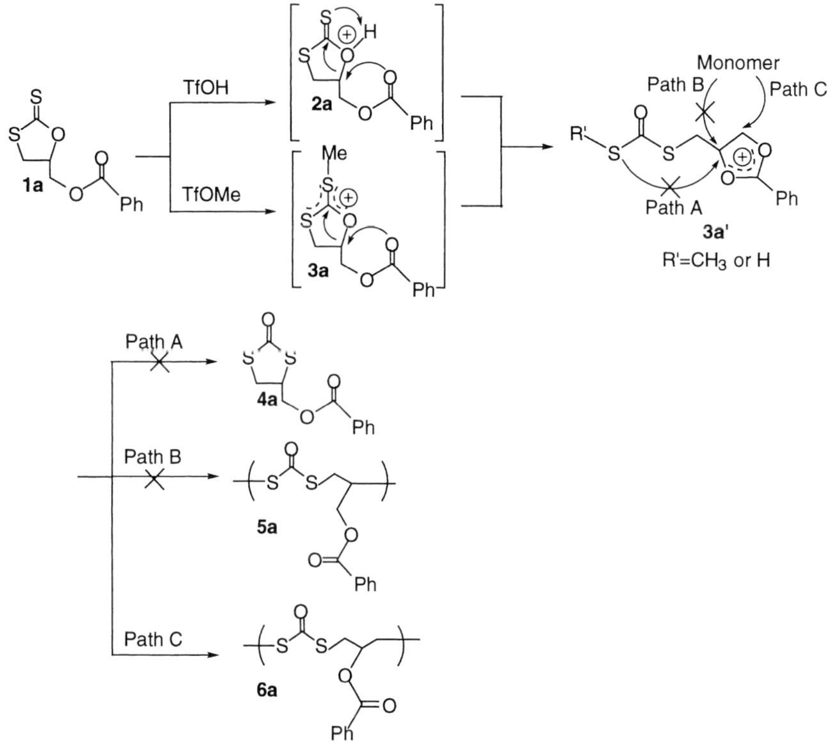

Conclusion

We have demonstrated the first example of a controlled living cationic ring-opening polymerization of the five-membered cyclic dithiocarbonate based on the neighboring group participation. This new concept of living polymerization may be applied to the design of well-defined novel functional polymers.

References

1. M. Szwarc, *Nature* **178**, 1168 (1956)
2. T. Aida, R. Mizuta, Y. Yosida, S. Inoue, *Makromol. Chem.* **182**, 1073 (1981)
3. L. R. Gilliom, R. H. Grubbs, *J. Am. Chem. Soc.* **108**, 733 (1986)
4. M. K. Georges, E. Rizzardo, P. M. Cacioli, G. K. Harmer, *Macromolecules* **26**, 2987 (1993)
5. M. Miyamoto, M. Sawamoto, T. Higashimura, *Macromolecules* **17**, 265 (1984).
6. R. Faust, J. P. Kennedy, *Polym. Bull.* **15**, 317 (1986)
7. R. Faust, J. P. Kennedy, *Polym. Bull.* **19**, 21 (1988)
8. T. Higashimura, Y. X. Deng, Y. X. *Polym. J.* **15**, 685 (1983)
9. D. Vofsi, A. V. Tobolsky, *J. Polym. Sci., Part A.* **3**, 3261 (1965)
10. S. Kobayashi, T. Saegusa, in: *Ring-Opening Polymerization*, Vol. 2, K. J. Ivin, and T. Saegusa (Eds.), Elsevier, London, 1989, p.761
11. W. Choi, F. Sanda, N. Kihara, T. Endo, *J. Polym. Sci., Part A: Polym. Chem.* **35**, 3853 (1997)
12. W. Choi, F. Sanda, T. Endo, *Macromolecules* **31**, 2454 (1998)
13. L. Goodman, *Adv, Carbohydrate Chem.* **22**, 109 (1967)
14. C. H. Fotsch, A. R. Chamberlin, *J. Org. Chem.* **56**, 4141 (1991)
15. N. Kihara, Y. Nakawaki, T. Endo, *J. Org. Chem.* **60**, 473 (1995)

SYNTHESIS OF CONDUCTING BLOCK AND GRAFT COPOLYMERS WITH POLYETHER SEGMENTS

Yusuf Yagci*[1] and Levent Toppare[2]

[1]Istanbul Technical University, Department of Chemistry, Maslak, Istanbul 80626, Turkey
[2]Middle East Technical University, Department of Chemistry, Ankara 06533, Turkey

Abstract: Synthesis of block and graft copolymers containing polyether and conducting polypyrrole sequences were described. Pyrrole moieties were incorporated at the chain ends of polytetrahydrofuran and polysiloxane and at the side chains of polyethyl vinylether by ionic polymerization and appropriate chemical reactions. Subsequent electropolymerization with pyrrole through these moities yielded free standing films of the corresponding block and graft copolymers. The formation copolymers was evidenced by FTIR spectroscopy and extraction with solvent of the precursor homopolymers. The thermal and morphological properties were also characterized. The two surface of the copolymer films generally differ in appearance, the surface at the solution side being cauliflowerlike or wrinkled, whereas the surface at the electrode side smooth. Conductivies of the copolymers were comparable with that of the pure polypyrrole.

INTRODUCTION

Conducting organic polymers has received growing interest because of their wide range applications in various fields [1]. These materials are often termed as synthetic metals due to the fact that they combine chemical and mechanical properties of the polymers with the electronic properties of the metals and semicunductors. Polypyrrole is one type of the polymer among a general type of conducting polymers that include polyacetylenes, polythiophenes, polyanilines, polyphenylenes, polycarbazoles, polyquinolines and polyphtalocyanines. Pyrrole systems have several attractiveness over the other conducting polymers. These include the chemical and thermal stabilty of the pyrrole based polymers and abilty of the preparation of derivatives with a range of conductivies. However, two major limtiations of polypyrroles are inabilty to process and poor mechanical and phsical properties. In order to improve their various properties several methods were proposed. Among them block and graft copolymerization is an elegant way to combine the conducting polypyrrole with the conventional polymers. For example, polystyrene-*g*-

CCC 1022-1360/00/$ 17.50+.50/0

polypyrrole [2] and poly(mehyl methacrylate)-*g*-polypyrrole [3-5] were prepared by electrochemical and combination of chemical and electrochemical methods, respectively.

In this paper, we will describe our approach to synthesize polypyrrole with soft polyether segments. In this approach, two mutually exclusive polymerization mechanisms, namely conventional ionic and electrochemical polymerizations, are sequentially combined. The elecroactive pyrolle moieties were introduced polymers by either termination of cationic polymerization by suitable nucleophile or chemical reaction of the related compunds with the preformed polymers The overall procedure is described below.

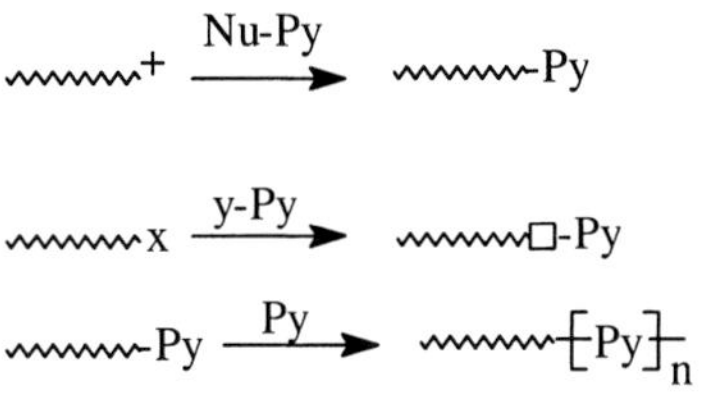

$$\text{wwww}^+ \xrightarrow{\text{Nu-Py}} \text{wwww-Py} \tag{1}$$

$$\text{wwww-x} \xrightarrow{\text{y-Py}} \text{wwww}\square\text{-Py} \tag{2}$$

$$\text{wwww-Py} \xrightarrow{\text{Py}} \text{wwww}\left[\text{Py}\right]_n \tag{3}$$

Py = pyrrole, Nu = nucleophile, x and y = mutually reactive functional groups

POLYPYRROLE/POLYTETRAHYDROFURAN BLOCK COPOLYMERS

Living polymerization allows the preparation of various well-defined polymers with functional end groups. It is known that tetrahydrofuran can be polymerized without chain transfer and termination reactions under closely controlled conditions [6]. The oxonium group of the living chain can react with nucleophiles and thus gives rise to a variety of functional groups [7-9]. Polytetrahydrofuran (PTHF) with one or two terminal pyrrole groups was prepared [10, 11] by the living polymerization of tetrahydrofuran (THF) using methyl triflate and triflic anhydride as mono- and bifunctional initiators, respectively, according to the following reaction:

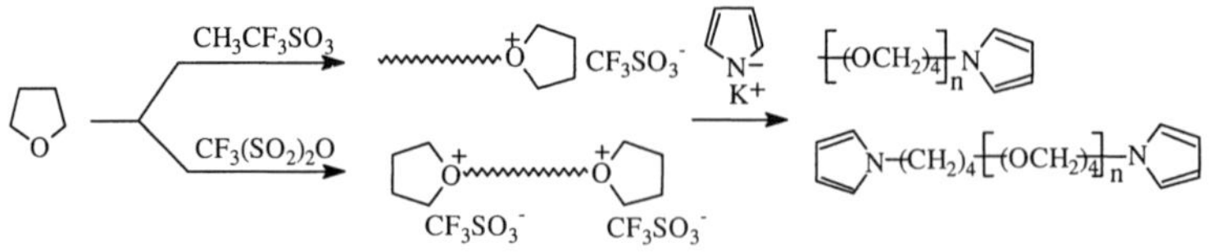

The living propagating chains readily react with potassium salt of pyrrole.The results are collected in Table 1. Based upon the comparision of the ratio of pyrrole protons at 6.4-7.1 ppm to protons of –OCH$_2$ of PTHF at 3.4 ppm, the functionalization of the polymers were calculated. UV spectra of the polymers show chracteristic absorbance of pyrrole moieties.

Table 1. Preparation of pyrrole terminated polytetrahydrofuran.

Polymer	Initiator (mol l^{-1})	Time (min)	Conv. (%)	M$_n$	M$_w$/M$_n$	F
Py-PTHF-1	CH$_3$CF$_3$SO$_3$ (1.2 x 10^{-2})	40	12	13000	1.5	1
Py-PTHF-2	CH$_3$CF$_3$SO$_3$ (1.2 x 10^{-2})	80	20	27000	1.7	1
Py-PTHF-Py-1	(CF$_3$SO$_2$)$_2$O (1.2 x 10^{-2})	20	10	30000	1.3	2.01
Py-PTHF-Py-2	(CF$_3$SO$_2$)$_2$O (2.44 x 10^{-2})	35	20	13800	1.5	1.98

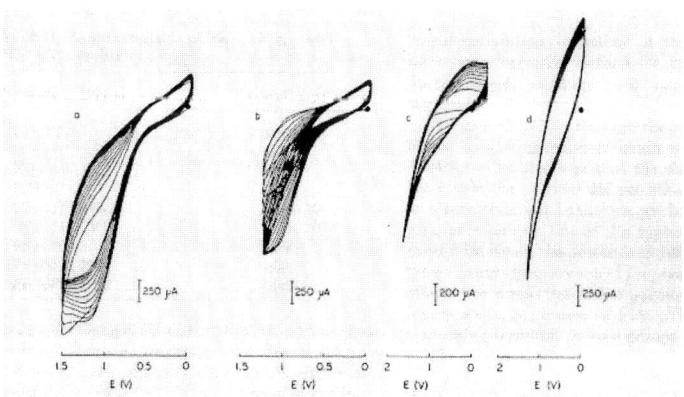

Figure1. Cyclic voltammograms of (a) TS$^-$ doped PPy, (b) TS$^-$ doped Py-PTHF-1, (c) BF$_4^-$ doped PPy and (d) BF$_4^-$ doped Py-PTHF.

The polymers obtained this way were not electroactive themselves as was confirmed by the cyclic voltammetry analyses. The absence of detection of any redox peak indicated that crosslinking through the pyrrole moieties was not possible. Different situation was, however, encountered when poyrrole was added to the solution. An electroactivity increasing with the number of scans was observed (Figure 1). Thus, these polymers were transformed into block copolymers containing pyrrole and THF segments by electropolymerization via single- or two-step procedures. In the single-step procedure, polymerizations were performed on a bare Pt electrode by using a conventional three-electrode electrochemical cell and BF_4^- or ClO_4^- as supporting electrolytes. Pyrrole, PTHF and electrolyte were dissolved in an appropriate solvent. Notably, sufficient amount of films were formed in much shorter times when ClO_4^- was used. In the two-step procedure, however, mono or bifunctional PTHFs were coated on the Pt electrode and subsequently pyrrole was polymerized electrochemically. Electrochemical synthesis of block copolymers yielded free standing films which were easily peeled off from the electrode surface.

The mechanical properties of block copolymers were quite different from the pyrrole homopolymer. The block copolymers produced rubbery films. These films were soft but retained their shape when cut into strips. FTIR spectra of the block copolymers showed characteristic bands of the precursor PTHF, polypyrrole segment and also the dopants. Thermogravimetric analyses revealed that the block copolymers doped with p-toluene sulfonate (TS⁻) were more resistant to heating when compared to BF_4^- and ClO_4^- in accordance with the thermal behaviour of pure polypyrrole. Block copolymers obtained from mono and bifunctional PTHF showed similar DSC paterns. Decomposition transitions of polypyrrole and the dopant were detected. Scanning electron microscopy (SEM) studies revealed that the surface appearance of the electrode and solution side depends on the type of electrolyte used. Smooth appearance in the electrode side and cauliflowerlike appearance, particularly with TS⁻, dopant indicates that pyrrole groups grew towards the solution.

In contrast to the chain length, the electrolyte type significantly changed the conductivities. The highest conductivity, very close to that of pure PPy, was shown by BF_4^- doped polymers. The lowest conductivity was shown by ClO_4^- doped films. When TS⁻ was used as the supporting electrolyte, films had conductivity values of about 0.07 S/cm, which was also the conductivity of of TS⁻ doped pure PPy (Table 2).

Table 2. Conductivities of the films doped with various dopants.

Polymer	Conductivity (S/cm)
PPy BF_4^-	0.56
PTHF-PPy-1 BF_4^-	0.56
PTHF-PPy-2 BF_4^-	0.14
PPy ClO_4^-	0.085
PTHF-PPy-1 ClO_4^-	5.5×10^{-3}
PTHF-PPy-2 ClO_4^-	9.2×10^{-3}
PPy TS⁻	0.09
PTHF-PPy-1 TS⁻	0.032
PTHF-PPy-2 TS⁻	0.065

POLYETHYLVINYLETHER/POLYPYRROLE GRAFT COPOLYMERS

Vinylethers are readily polymerized by onium type photoinitiators. Both direct and indirect acting systems were found to be effective for the initiation [12]. Among various indirect acting systems radical promoted cationic polymerization is particularly useful and flexible way for the polymerization of vinyl ethers [13]. The operating wavelenght is extended to longer wavelenghts with the aid of free radical photoinitiators so that the absorption of the onium salt does not interfere with that of the chromophoric group of a particular monomer involved. For convenience, we have employed, 2,2-dimethoxy-2-phenyl acetophenone (DMPA) / diphenyl iodonium hexafluoro phosphate ($Ph_2I^+PF_6^-$) combination for the radical promoted cationic polymerization of 2-choloroethyl vinly ether (CEVE). Irradiation of solution of

CEVE (6.6 mol l^{-1}) in methylene chloride containing DMPA (5 x 10^{-3} mol l^{-1}) and $Ph_2I^+PF_6^-$ (5 x 10^{-3} mol l^{-1}) at λ = 350 nm for 5 minutes produced a polymer with 98 % conversion. In this case the initiating species are formed by the oxidation of either or both primary and monomer adduct radicals.

$$(8)$$

$$(9)$$

$$(10)$$

$$(11)$$

PCEVE thus obtained was converted to pyrrole derivative (polypyrrolethyl vinyl ether, PPEVE) by reacting with pyrroyl potassium in THF in the presence of 18-crown-6.

$$(12)$$

Structure of PPEVE was confirmed by IR, ^{1}H-NMR and ^{13}C-NMR spectral and thermal analyses. Glass transition temperature of PCEVE increased from –6.67 $^{\circ}$C to 4.13 $^{\circ}$C by such modification indicating that PPEVE is relatively more rigid than the corresponding prepolymer. Moreover, subtitution increases the thermal stability as demonstrated with TGA studies. Electropolymerization of pyrrole with PPEVE was performed in both in organic solution and water as described for PTHF.

$$\left[CH_2-\underset{\underset{CH_2CH_2-N}{\overset{|}{O}}}{CH}\right]_n \xrightarrow[\text{Electro polym.}]{} \left[CH_2-\underset{\underset{CH_2CH_2-N}{\overset{|}{O}}}{CH}\right]_n \quad \text{(pyrrole units)} \tag{13}$$

Although the chemical incorporation of polyether with polypyrrole in both cases was evidenced by the treatment of the graft copolymers with solvents and IR spectral analysis, DSC studies revealed that electropolymerization in organic solution yielded graft copolymers with higher pyrrole contents. The morpholoy of the graft copolymers were different than that of the pure PP. Electrode side of the pure PPy produced under the same experimental conditions exhibit wrinkled surface whereas the corresponding surface of the graft copolymer is smooth [14].

No weight loss is recorded upon washing the graft copolymers with the solvent of PPEVE. Conductivities of washed and unwashed graft copolymer films were almost the same.

POLYSILOXANE / POLYPYRROLE BLOCK COPOLYMERS

Excellent properties of polysiloxanes regarding high flexibility, high temperature resistance and solubility prompted us to modify the properties of polypyrrole by similar electrochemical blocking process of polysiloxane with pyrrole. Many synthetic approaches for siloxane containing block and graft copolymers are based on organofunctional polysiloxanes as the starting materials [15]. Organofunctional polysiloxanes can be prepared [16] by several routes such as hydrosilation, chemical transformation and the cationic or anionic equilibration of cyclic siloxanes in the presence of a functional disiloxane acting as an end-blocking agent. Pyrrole functionalized polydimethylsiloxane (Py-PDMS-Py) was prepared by a multistep procedure according to the following reactions.

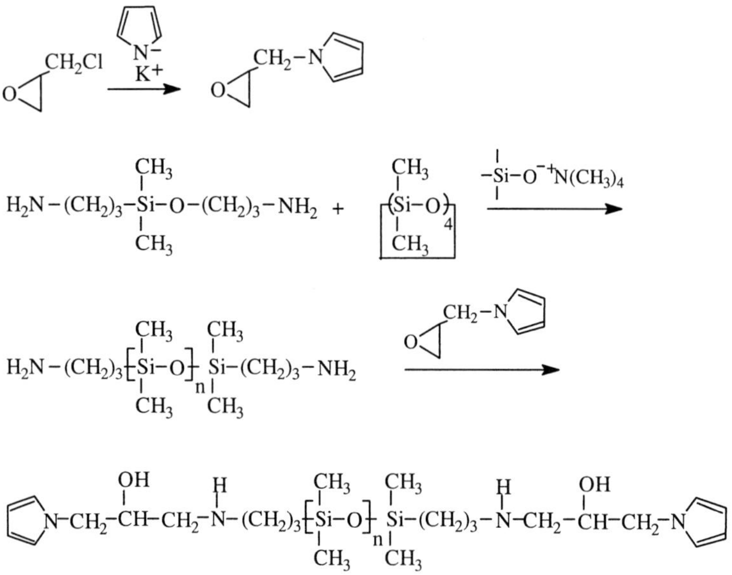

$$\text{(14)}$$

$$\text{(15)}$$

$$\text{(16)}$$

In this procedure, the functionalization agent N-glycidylpyrrole was prepared by reacting pyrrole salt with epichlorphydrin in the presence of 18-crown-6 in THF. The reaction of amino compounds with epoxy-terminated polysiloxanes is a known process for polysiloxane functionalization. For this purpose, the amino groups were attached to polysiloxane chain by anionic polymerization of octamethyl-cyclotetrasiloxane in the presence of 1,3-bis(aminopropyl)-1,1,3,3-tetramethyl-siloxane as the end-blocking agent. The good agreement of the molecular weight values obtained by titration of the amino end groups ($M_n=1100$) and by vapour pressure osmometry ($M_n=1210$) confirms the presence of two functional groups per polymer chain and the absence of important amounts of high molecular weight cyclosiloxanes. In the subsequent step, this polymer was reacted with N-

glycidylpyrrole to yield pyrrole terminated polydimethylsiloxane (Py-PDMS-Py). The molecular weight of Py-PDMS-Py (M_n=1200) was calculated from the pyrrole content, as determined from the ^{1}H-NMR spectrum, and was found to be in good agreement with that of the amino derivative.

The block copolymers were then prepared by electrochemical polymerization of pyrrole on the surface of the electrode coated with Py-PDMS-Py.

(17)

The block copolymers doped with TS$^-$ and BF$_4^-$ were investigated by FTIR analysis. The characteristic Si-O-Si bands of the polysiloxane and C=C bands of polypyrrole at 1096 and 1023 cm^{-1} and 1550 cm^{-1}, respectively, were detected. Thermal behaviour of the block copolymer as studied by DSC and TGA depended on the type of dopant used. Although the size of the films was not suitable for the detailed tests, the electrolytic films were subjected to simple tests. They were foldable without any cracks on the surface.

The morphology of the films, as judged by SEM, was dependent of the electrode or solution side. The type of the dopant was shown to be also important determinant of the appearance of the surfaces. The electrode side of the block copolymer doped with TS$^-$ was much smoother than that of the block copolymed doped with BF$_4^-$.

Conductivities of the solution and electrode sides of the block copolymer films were found to be the same. Conductivities of the block copolymers (doped either with TS$^-$ and BF$_4^-$) were observed in the same order of magnitude (2-5 S/cm). This indicates the homogenity of free standing films.

In conclusion, main chain and side chain pyrrole functionalized polyethers have been successfully synthesized by ionic polymerization processes and chemical reactions. Block and graft copolymer films of the polyethers and polypyrrole were futher synthesized by electrochemical methods.

38

ACKNOWLEDGEMENTS

The authors would like to thank DPT (DPT-95K 120498) and Istanbul Technical University Funds for their financial support.

REFERENCES

1. T.A. Skotheim, Ed., "Handbook of Conducting Polymers" Marcel Dekker, New York, 1986

2. A.I. Nazzal, G.B. Street, J.Chem.Soc., Chem. Commun., 375 (1985)

3. M.L. Hallensleben, D. Stanke, Macromol.Chem. & Phys. 75, 190 (1995)

4. D. Stanke, M.L. Hallensleben, L. Toppare, Synth.Met., 55-57, 1108 (1993)

5. D. Stanke, M.L. Hallensleben, L. Toppare, Synth.Met., 73, 261 (1995)

6. F. D'Haese, E. J. Goethals, Br. Polym. J., 20, 103 (1988)

7. Y. Tezuka, E.J. Goethals, Eur.Polym.J., 18, 991 (1982)

8. G.Hizal, Y.Yagci, W.Schnabel, Polymer, 20, 4443 (1994)

9. A.Onen, Y.Yagci, Angew. Makromol. Chem., 243, 143 (1996)

10. N. Kizilyar, L. Toppare, A. Onen, Y. Yagci, Polym. Bull., 40, 639 (1998)

11. N. Kizilyar, L. Toppare, A. Onen, Y. Yagci, J. Appl. Polym. Sci., 71, 713 (1999)

12. Y.Yagci, I. Reetz, Prog. Polym.Sci., 23, 1485 (1998)

13. Y.Yagci, W. Schnabel, Makromol. Chem., Macromol. Symp., 60, 133 (1992)

14. E. Kalycioglu, L. Toppare, Y.Yagci, submitted to Synth. Met.

15. I.Yilgor, J.E. Mcgrath, Adv. Polym.Sci., 86, 1, (1988)

16. V. Harabagiu, M. Pintela, C. Cotzur, B.C. Simonescu, in Polymeric Materials Encylopedia: Synthesis, Properties and Applications, Vol. 4, CRC Press, Boca Raton, Fl, 1996, p. 2661

17. E. Kalaycioglu, L. Toppare, Y. Yagci, V. Harabagiu, M. Pintela, R. Ardelen, B.C. Simionescu, Synt. Met., 97, 7 (1998)

CONTROLLED RING-OPENING POLYMERIZATION OF LACTONES AND LACTIDES

Ann-Christine Albertsson,* Ulrica Edlund and Kajsa Stridsberg

Department of Polymer Technology, The Royal Institute of Technology

S-100 44 Stockholm, Sweden, e-mail: aila@polymer.kth.se

SUMMARY: 1,5-dioxepan-2-one (DXO) is presented as a versatile component in biodegradable polymers for biomedical applications. Copolymerization of DXO and L-lactide yielded a semi-crystalline, yet flexible, material where the extent of crystallinity and erosion characteristics were controlled by an appropriate choice of copolymer composition. Crosslinked PDXO was polymerized as a novel biodegradable elastomer. The degradation behavior of these materials were explored *in vitro*. Microspheres from poly(DXO-co-L-LA) were prepared and shown to be promising candidates for controlled release. The polymer composition and drug solubility provided effective means of controlling the drug delivery pattern.

Introduction

Biomaterials are defined as a non-viable material, used in a medical device for the replacement of living material, to repair, restore, or replace damaged or diseased tissue or to interact with biological systems. The use of non-sustainable materials, degradable by hydrolysis into products resorbed by the body with a minimal tissue response, minimizes the number of surgical operations. Degradable synthetic polymers such as aliphatic polyesters, especially those derived from lactide (LA), glycolide (GA), and polyanhydrides have been thoroughly investigated [1,2]. Both homo- and copolymers of synthetic poly(hydroxy acid)s are important degradable materials on the market today due to adjustable degradation rate and at the same time good initial mechanical properties. This study presents 1,5-dioxepan-2-one (DXO), as a component in L-lactide copolymers or as a crosslinked biodegradable elastomer, as versatile and very promising bioresorbable materials in the field of surgery and pharmaceuticals.

Over the years we have developed the controlled ring-opening polymerization (ROP) of lactones, lactides and cyclic anhydrides to prepare polymers where the elasticity, degradability and biocompatibility are regulated. The ROP provides a versatile way to achieve pure and well defined structures. Homopolymers, random and crosslinked copolymers with different complex molecular architecture are prepared using ROP with multifunctional initiators, crosslinkers or termination agents[3]. The work involves the synthesis of monomers, studies of the polymerization kinetics and mechanism[4], and characterization of the polymers to form materials suitable for pharmaceutical applications such as drug delivery[5], or as surgical devices like sutures.

The interest and use of biodegradable polymers in controlled drug delivery systems is constantly growing. The incorporation of therapeutic agents in a polymeric matrix serves two main purposes; the protection of the drug from physiological degradation and the sustained release of drug to the patient. Microspheres from poly(DXO-co-L-LA) were prepared by a solvent evaporation technique. The polymer composition, molecular weight and stereochemistry offer important tools to tailor-make different release profiles for specific applications.

Experimental

Polymerization. Copolymers of 1,5-dioxepan-2-one and L-lactide were prepared by bulk copolymerization with stannous octoate as catalysts at 120°C[6]. After the desired reaction time, the reaction vessel was cooled and the content was dissolved in chloroform and precipitated in cold methanol. Crosslinked poly(1,5-dioxepan-2-one) (X-PDXO) was prepared by transesterification between the poly(DXO) and the crosslinker 2,2-Bis((ε-caprolactone-4-yl)-propane) (BCP) or bis(ε-caprolactone-4-yl) (BCY)[3]. Transesterification was conducted in bulk at 140°C with stannous octoate as catalyst.

In vitro **hydrolysis.** Hydrolysis studies were performed with melt-pressed films of the copolymers, 0.5 mm in thickness. Circular disks with a diameter of 13 mm were immersed in a phosphate buffer of pH=7.4. Degradation was carried out at 37°C without shaking or stirring motions. *In vitro* studies of the X-PDXO were conducted with polymer films or rods and agitated gently on a mechanical shaker under otherwise equal conditions.

Preparation of microspheres. Microspheres from poly(DXO-co-L-LA) having a DXO:L-LA ratio of 30:70 and 15:85 respectively were prepared and drugs were incorporated using an oil-in-water solvent evaporation technique. A homogenous solution of polymer and drug in methylene chloride was added dropwise to a water phase containing 1% (w/v) poly(vinyl alcohol) under vigorous stirring. Dropsize was reduced by sonication. After precipitation at ambient temperature, the microspheres were recovered by centrifugation, washing in destilled water and drying under vacuum. A hydrophilic (timolol maleate) and a hydrophobic (amitryptiline) model drug were studied.

***In vitro* degradation and drug release.** Degradation and drug release studies were conducted at 37 °C by immersing pre-weighed samples in 0.1 M phosphate-buffered saline (pH 7.4). Duplicate samples were removed at selected times and the microspheres were isolated by filtration on pre-weighed AcetatePlus supported filters, dried under vacuum and analyzed. Mean values were calculated and used. The saline buffer was periodically analyzed for the release of drug by measuring the extinction at 294 nm using a 8451A Diode Array UV-VIS Spectrophotometer from Hewlett Packard.

Instrumentation. ^{1}H-NMR spectra were obtained by a Bruker Avance DMX 500 NMR spectrometer. Molecular weight was monitored by size exclusion chromotography using a Waters equipment consisting of a Waters 6000A pump with five Ultrastyralgel® columns (10^5, 10^4, 10^3, 500 and 100Å pore sizes) and a Polymer Labs evaporative light scattering detector. Chloroform was used as eluent and polystyrene standards with narrow molecular weight distributions (M_w/M_n=1.06) were used to calibrate the system. DSC thermograms were recorded on a Mettler Toledo DSC 820 connected to a RP100 cooling unit from Labplant, England. A JEOL JSM 5400 scanning electron microscope was employed to examine the surface morphology of films. Samples were mounted on metal stubs and sputter coated with gold-palladium (Denton Vacuum Desc II).

Results and discussion

Polymerization. Extensive research is performed by our research group to develop degradable polymers with properties that can be altered for different applications. The materials obtained have been shown to exhibit adjustable properties by different kinds of modifications, by copolymerization, crosslinking or surface treatment. The copolymerization between DXO and L-LA was conducted to produce materials suitable for drug delivery.

Due to the difference in reactivity ratio for the two monomers, $r_{(L\text{-lactide})}=10$ and $r_{(DXO)}=0.1$, the polymer formed was not a true random copolymer but rather a semi-block copolymer.

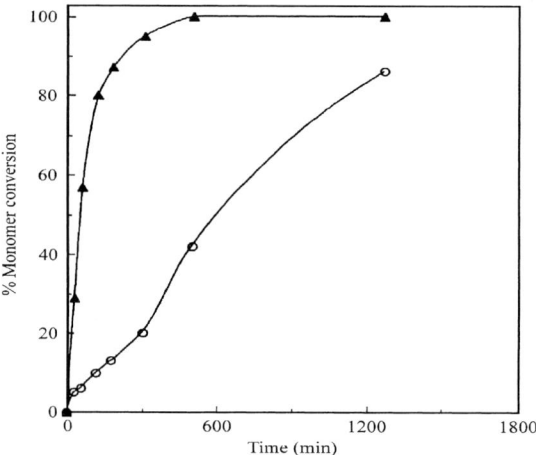

Figure 1. The monomer conversion as a function of the reaction time for the copolymerization of L-lactide (▲) and 1,5-dioxepan-2-one (O).

Figure 1 shows that the L-lactide monomer was consumed predominantly during the first part of the polymerization, the DXO monomer was incorporated in the macromolecule during the later stages of reaction. The copolymers were analyzed with NMR spectroscopy, which revealed that the polymer composition were in agreement with the feed ratios. ^{13}C-NMR studies showed that transesterification reactions had taken place and redistributed the monomer sequence. Only shorter homopolymer segments with an average length of about 3 to 5 units could be detected. This feature made the resulting material semi-crystalline, exhibiting thermal properties that could be adjusted by changing the polymer composition. Figure 2 shows the crystallinity as a function of the amount of DXO. The crystallinity decreases with increasing amount of DXO up to a DXO content of 40% after which no crystallinity is detectable.

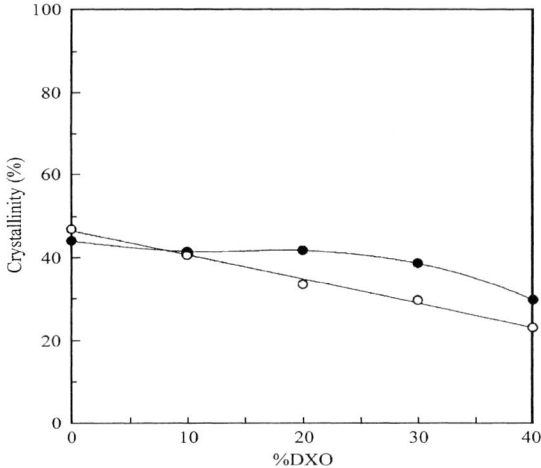

Figure 2. The crystallinity as a function of the amount of DXO in the copolymer. Open circles denote crystallinity determined by DSC, filled circles crystallinity determined by X-ray diffraction.

Microspheres, *in vitro* drug delivery A oil-in-water solvent evaporation technique was employed to prepare microspheres with encapsulated drug from poly(DXO-co-L-LA)[5]. Figure 3 shows that the obtained particles were smooth and dense without visible pores on the sphere surface. The average diameter was 65 μm. The internal structure was slightly polarized with a greater porosity in the bulk than on the external face.

Molecular weight loss was apparent immediately after incubation, whereas the mass loss was characterized by a lag period of 2 weeks followed by a gradual loss of material. This degradation pattern is typical of a bulk eroding polymer and well consistent with the classical theory of polyester hydrolysis.

44

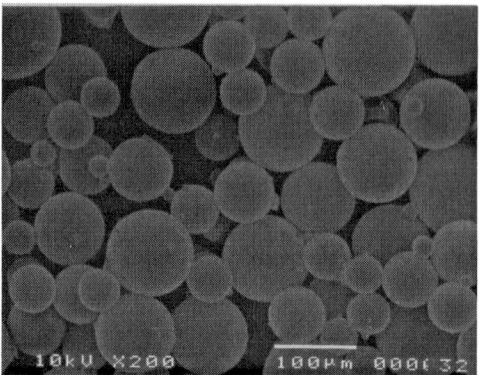

Figure 3. Scanning electron micrograph of poly(DXO-co-L-LA) microspheres.

^{1}H-NMR analysis of copolymer composition revealed the enrichment of lactide units over degradation time. The crystalline lactide regions were more resistant to water penetration and diffusion and were thus preserved in the structure although being more hydrolytically susceptible than the amorphous DXO units.

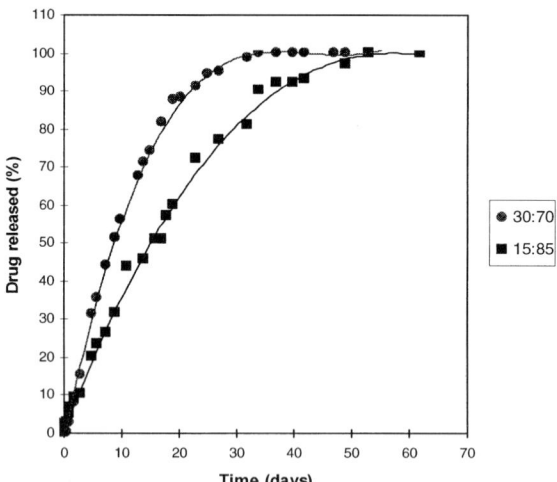

Figure 4. The release of TM from (●) poly(DXO-co-L-LA)(30:70) and (■) poly(DXO-co-L-LA)(15:85) microspheres.

Whilst a hydrophilic substance was released within 4 weeks, a hydrophobic drug required up to 3 months for depletion. In the former case, diffusion was the dominating release mechanism whereas the release of hydrophilic substance was mainly erosion controlled. Sustained release of an incorporated hydrophilic drug was obtained, as shown in Figure 4. The difference in morphology between copolymers with different DXO:L-LA ratios had a marked effect upon the rate of drug delivery, thus providing an excellent tool for controlling the release behavior.

Crosslinked poly(1,5-dioxepan-2-one). A novel polymer under evaluation as a potential material for medical applications is the crosslinked poly(1,5-dioxepan-2-one). Studies showed that crosslinked poly(DXO) was easily formed by exposing the poly(DXO) to stannous octoate and crosslinker at 140°C[3] (Scheme 1). The formed material were elastic and totally amorphous with a Tg of –35°C for the BCP crosslinked films and –21°C for the BCY crosslinked ones. Analysis of the molecular weight between the crosslinks by swelling experiments, indicated that homogeneous crosslinking had taken place.

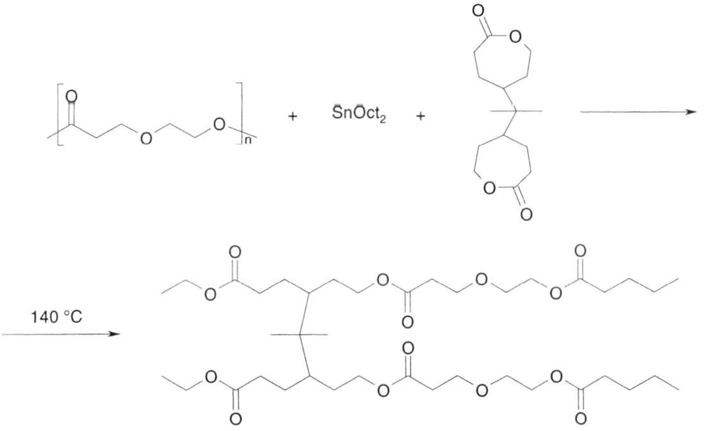

Scheme 1. Formation of crosslinked poly(DXO) by transesterification between poly(DXO) and BCP at 140°C with stannous octoate as catalyst.

The material was subjected to various kinds of surface treatments, e.g. electron beam irradiation and surface grafting with acryl amide. The degradation of crosslinked and modified poly(DXO) is shown in Figure 5. During the initial stages of hydrolysis the untreated samples showed a bigger mass loss than the irradiated and surface grafted.

Both the irradiated and the surface grafted samples exhibited a plateau in mass loss up to 16 weeks after which the rate of degradation was comparable for all samples. A total weight loss was achieved after about 300 to 350 days *in vitro*.

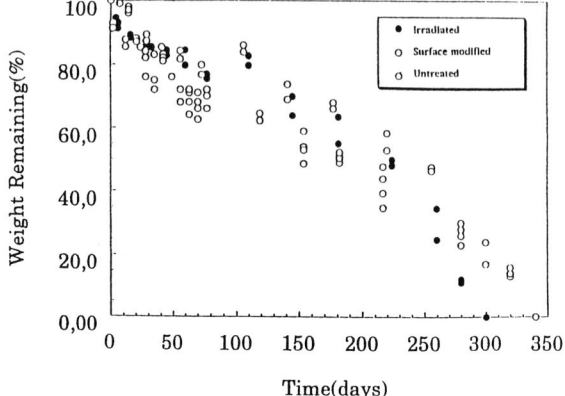

Figure 5. Weight loss in crosslinked poly(DXO) with a crosslinking density of 16% BCP, degradation in saline buffer 37°C.

Conclusions

It is possible to prepare a wide range of materials, varying from physically crosslinked copolymers of L-lactide and DXO, to strong and tough crosslinked poly(DXO) networks by ring-opening polymerization. Copolymerization of L-lactide and 1,5-dioxepan-2-one were studied and the material formed was used as a novel matrix for sustained drug delivery. Smooth and dense microspheres were successfully prepared by an oil-in-water technique. Sustained release of incorporated drugs was obtained. The polymer composition and the drug solubility were proven effective instruments of modifying the rate of hydrolytically degradation, erosion, and drug release.

References

[1] Albertsson A.-C.; Karlsson S. Encyclopedia of Environmental Analysis and Remediation, R.A. Meyers Ed., John Wiley & Sons, Inc. 1998
[2] Albertsson A.-C.; Eklund M. J. Appl. Polym. Sci.1995, 57, 87
[3] Palmgren, R; Karlsson, S; Albertsson, A.-C. J Polym Sci Polym Chem, 1997, 35, 1635
[4] Stridsberg, K.; Albertsson, A.-C. J Polym Sci Polym Chem, in press
[5] Edlund, U.; Albertsson, A.-C. J Polym Sci Polym Sci, 1999, 37, 1877
[6] Albertsson, A.-C.; Löfgren, A. J Macromol Sci Chem, 1995, A32, 41

Novel functionalization routes of poly(ε-caprolactone)

Ph. Lecomte, Ch. Detrembleur, X. Lou, M. Mazza, O. Halleux, R. Jérôme*

Center for Education and Research on Macromolecules (CERM), University of Liege, Sart-Tilman, B6, 4000 Liege, Belgium

Abstract: The aluminum alkoxide mediated ring opening polymerization of functional lactones, such as γ-ethylene ketal-ε–caprolactone (TOSUO), γ-(triethylsilyloxy)-ε–caprolactone (SCL) and γ–bromo-ε–caprolactone (γBrCL), is a versatile route to polyesters containing ketal, ketone, alcohol and bromide groups. As result of living polyaddition mechanism, random and block copolymerization of εCL and γBrCL has been successfully carried out. The reactivity ratios are quite similar (1.08 for ε-CL, and 1.12 for γBrCL). These random copolymers are semicrystalline when they contain less than 30 mol% of γBrCL, otherwise they are amorphous. No transesterification reaction occurs during the sequential polymerization of ε-CL and γBrCL leading to block copolymers. Reaction of poly(εCL-co-γBrCL) with pyridine provides quantitatively a polycationic polyester. Furthermore, the reaction of this random copolymer with 1,8-diazabicyclo[5.4.0] undec-7-ene (DBU) is a route to unsaturated polyesters, whose the non conjugated double bonds can be quantitatively converted into epoxides by reaction with m-chloroperbenzoic acid (mCPBA). No chain degradation is detected during these derivatization reactions of poly(εCL-co-γBrCL).

Introduction

Nowadays, a steadily increasing attention is paid to biodegradable polyesters for their potential in agriculture, waste management of plastics, medicine and surgery.[1]

Over the last 20 years, CERM has been involved in a research program dealing with the macromolecular engineering of polylactones and polylactides.[2] Aluminum alkoxides [Al(OCH$_2$CH$_2$X)$_3$] carrying a functional group (X=Br, CH$_2$NEt$_2$, CH$_2$CH=CH$_2$, OCOC(Me)=CH$_2$,...) or not (X = H) have been used as efficient initiators for the ring opening polymerization (ROP) of lactones, lactides, glycolide and cyclic anhydrides. Hydrolysis of the active aluminum alkoxide bond leads to the formation of an asymmetric telechelic polyester, the end groups being X and OH, respectively. The mechanism proceeds through the coordination of aluminum to the exocyclic carbonyl oxygen of ε-caprolactone (εCL), followed by the acyl-oxygen cleavage of the monomer and insertion into the Al-O bond of the initiator (Scheme 1). These polymerizations are living and they have been exploited for the macromolecular engineering of polyesters and polyanhydrides, including the tailoring of block copolymers and more complex architectures.

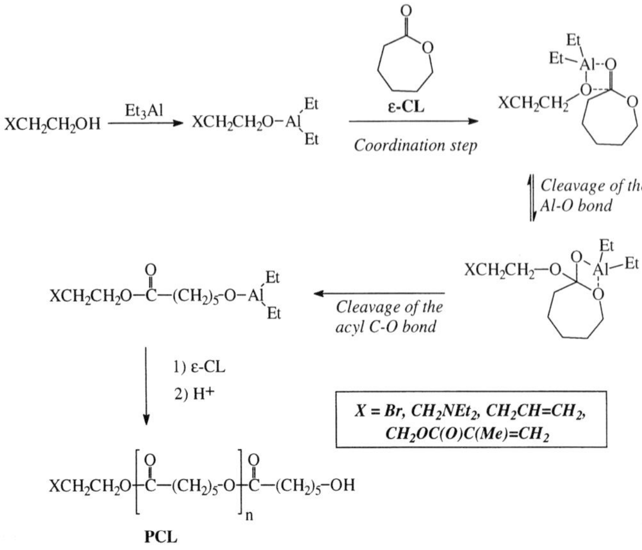

Scheme 1 : Mechanism for the aluminum alkoxide mediated ring opening polymerization of εCL.

Poly(ε–caprolactone) (PCL) shows a remarkable sets of characteristic features, such as biocompatiblity, permeability and miscibility with a wide range of polymers, e.g., PVC bisphenol A polycarbonate. Nevertheless, depending of the envisioned applications, some properties, such as hydrophilicity, biodegradation rate, and mechanical properties, have to be optimized. The εCL copolymerization with glycolide and lactides is a first tool to tailor the PCL properties. An alternative pathway relies on the synthesis and (co)polymerization of functional derivatives of εCL, so leading to polymers with functional pendent groups which are highly desirable for attaching drugs, tuning biodegradation rate, improving biocompatibility and promoting bioadhesion. The functional groups have to be protected or chosen in such a way that they do not interfere with the polymerization mechanism. A few years ago, synthesis of a few functional εCL's was known, although it was usually tedious and the polymerization was ill-controlled. [3]

γ-ethylene ketal-ε–caprolactone and γ-silyloxy-ε–caprolactone

More recently, we reported on the synthesis of γ-ethylene ketal-ε–caprolactone (TOSUO). [4,3] The TOSUO polymerization was initiated by aluminum alkoxide in toluene at 25°C and found to be living, which allowed for the controlled synthesis of poly(TOSUO-b-εCL) copolymers of narrow molecular weight distribution (Scheme 2). [5] No transesterification reaction occurred as proved by ^{13}C-NMR analysis. The end-group analysis agreed with the coordination-insertion mechanism previously reported for the εCL polymerization (Scheme 1). Random copolymerization of εCL and TOSUO was also carried out, and the reactivity ratios were measured by ^{13}C-NMR, being 1.3 for εCL and 1.0 for TOSUO. [6]

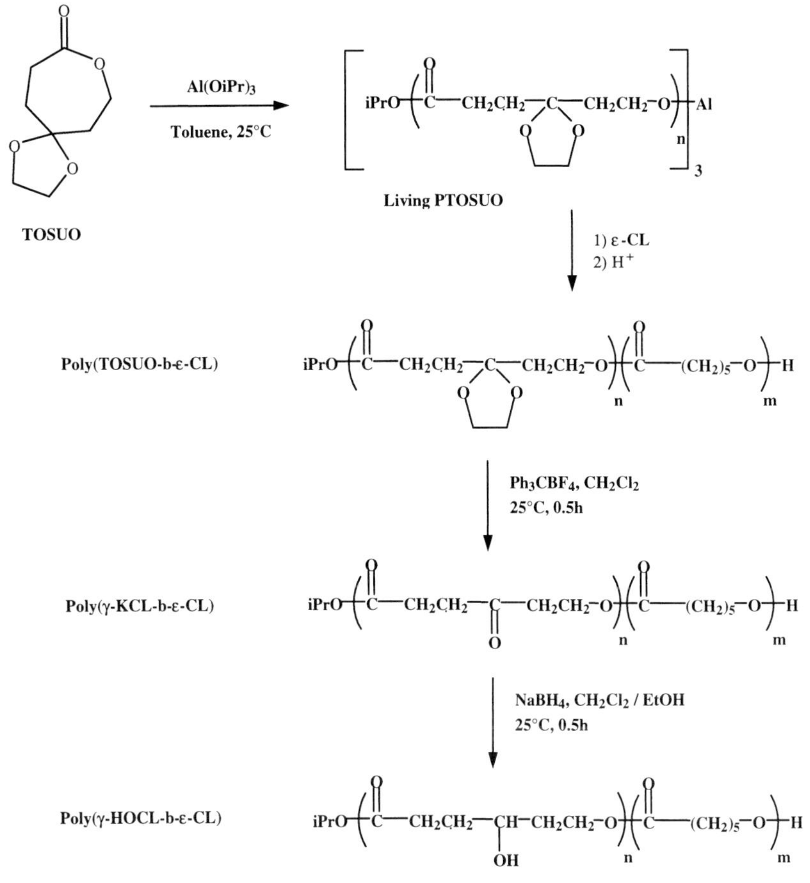

Scheme 2 : Synthesis of poly(εCL-b-TOSUO) copolymer and deprotection of the ethylene ketal pendent groups into ketones and alcohols, respectively.

The ethylene ketal pendent groups of the PTOSUO block were successfully deacetalized into ketones, which were further reduced into hydroxyl groups (Scheme 2). Therefore, PCL with well-defined content of ketone and hydroxyl pendent groups could be made available. It is worth pointing out that poly(2-oxepane-1,5-dione) (poly γ–KCL) prepared by deprotection of the homo PTOSUO is a highly crystalline modified poly(ε–caprolactone) with a high melting temperature of ca. 150°C.[7] Remarkably, the pendent hydroxyl groups can be reacted with triethylaluminum with formation of macroinitiators for ROP of lactones, lactides and glycolide. Complex molecular architectures, e. g., comb-shaped[4], graft[4], hyperbranched[8] and dendrigraft[9] polyesters, were accordingly synthesized.

An alternative pathway to polyesters containing hydroxyl pendent groups was also investigated based on γ-(t-butyldimethylsilyloxy)-ε-caprolactone (SCL)[10] and γ-(triethylsilyloxy)-ε–caprolactone (TeSCL)[11]. For instance, the poly(εCL-co-TOSUO-co-TeSCL) terpolymer was prepared with a predictable molecular weight and a narrow molecular weight distribution. The silanolate groups were selectively deprotected into hydroxyl groups, then converted into aluminum alkoxides, and the macroinitiator was used to synthesize graft copolymer. The subsequent conversion of the ethylene ketal groups into aluminum alkoxides followed by lactide ROP proved to be a valuable route to hetero-graft aliphatic copolyesters, as shown below (Scheme 3).[11]

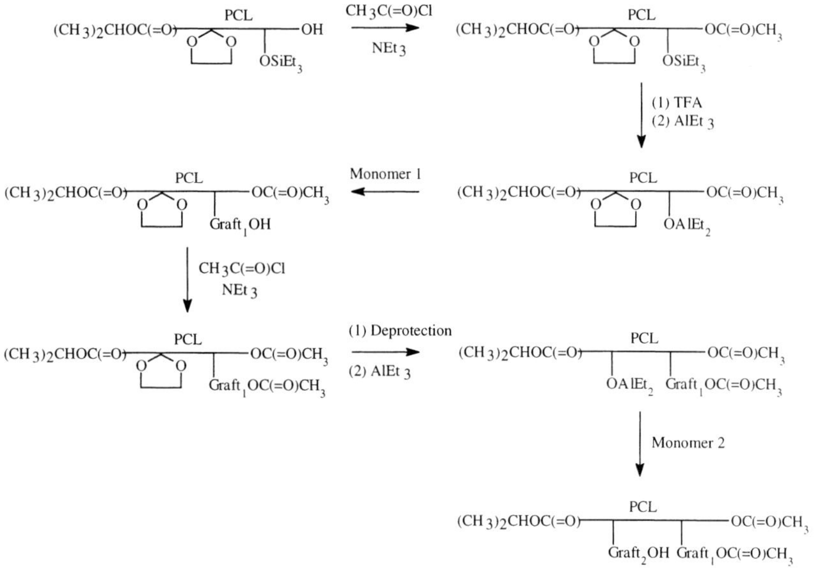

Scheme 3 : Synthesis of hetero-graft polyester from poly(εCL-co-TOSUO-co-TeSCL).

γ–Bromo-ε–caprolactone

Since bromine containing compounds are easily converted into a wide range of organic functions by well-known reactions of organic chemistry, e. g. elimination and substitution reactions, we have been interested in the synthesis and polymerization of γ–bromo-ε–caprolactone (γBrCL). The synthesis of this monomer has been reported elsewhere.[12] The γBrCL polymerization has been initiated by Al(OiPr)$_3$ in toluene at 0°C. All the molecular characteristic features of the polymer agree with a living polymerization process according to the same coordination-insertion mechanism as εCL and TOSUO. Differential Scanning Calorimetry (DSC) shows that PγBrCL is an amorphous polyester with a Tg of -16.5°C, which makes it comparable to PTOSUO (Tg = -11°C) but different from PCL which is typically semi-crystalline (Tg = -60°C and Tm=60°C).

Although the copolymerization of γBrCL and εCL mixtures was reported, the structure of the copolymer was not investigated in detail. A series of random copolymers of different composition has been prepared by changing the molar fraction of γBrCL in the comonomer feed (f$_B$), as shown in Table 1. The molar fraction of γBrCL (F$_B$) in the copolymer and the number-average molecular weight (M$_{n, H-NMR}$) have been measured by ^{1}H-NMR and found in agreement with the expected values f$_B$ and M$_{n,th}$. The molecular weight distribution is narrow in line with a living polymerization.

Table 1. Molecular characteristics of random copolymers of εCL and γBrCL.[a]

f$_B$[b]	F$_{B, H-NMR}$[c]	Conv. (%)	M$_{n, th}$[d]	M$_{n, H-NMR}$[l]	M$_{n,SEC}$[e]	M$_w$ / M$_n$
0.1	0.09	100	17,000	17,000	25,500	1.17
0.3	0.28	100	17,000	16,500	23,500	1.16
0.5	0.49	100	17,000	16,000	20,700	1.18

a) Conditions: toluene, 2h, 0°C
b) molar fraction of γBrCL in the comonomer feed
c) molar fraction of γBrCL in the random copolymer
d) M$_{n, th}$ = (114 * [εCL]$_0$ / [Al(OiPr)$_3$]$_0$) + (193 * [γBrCL]$_0$ / [Al(OiPr)$_3$]$_0$)
e) M$_{n,SEC}$ based on calibration valid to polystyrene.

As reported in the scientific literature, the carbonyl resonances observed by ^{13}C-NMR of the aliphatic polyesters are sensitive to the sequence effect (Scheme 4).[3]

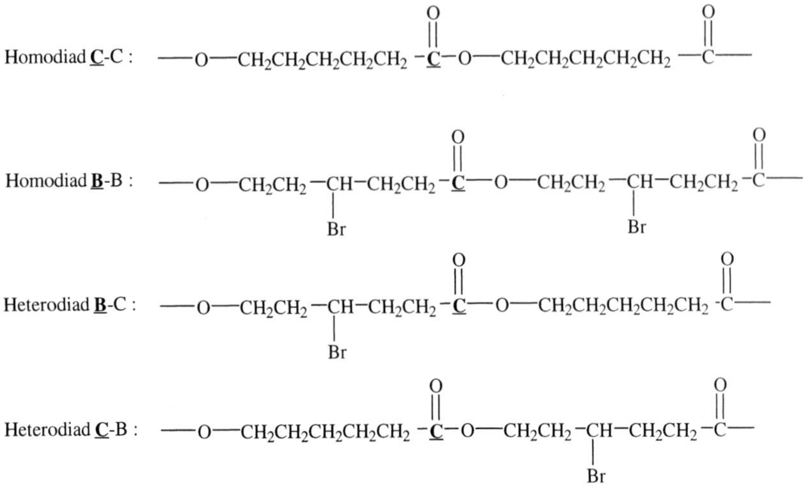

Homodiad **C**-C : —O—CH$_2$CH$_2$CH$_2$CH$_2$CH$_2$ —**C**—O—CH$_2$CH$_2$CH$_2$CH$_2$CH$_2$ —C—

$$\overset{O}{\overset{||}{}} \qquad \overset{O}{\overset{||}{}}$$

Homodiad **B**-B : —O—CH$_2$CH$_2$ —CH—CH$_2$CH$_2$ —**C**—O—CH$_2$CH$_2$ —CH—CH$_2$CH$_2$ —C—
                                      | |
                                      Br   Br

Heterodiad **B**-C : —O—CH$_2$CH$_2$ —CH—CH$_2$CH$_2$ —**C**—O—CH$_2$CH$_2$CH$_2$CH$_2$CH$_2$ ·C—
                                       |
                                       Br

Heterodiad **C**-B : —O—CH$_2$CH$_2$CH$_2$CH$_2$CH$_2$ —**C**—O—CH$_2$CH$_2$ —CH—CH$_2$CH$_2$ —C—
                                                                          |
                                                                          Br

Scheme 4 : Homodiads and heterodiads observed by ^{13}C-NMR.

Figure 1 shows the spectra of the carbonyl region for the homopolyesters and the random copolyester. Although only one resonance is observed for the carbon of the carbonyl in PCL (Figure 1a) and PγBrCL (Figure 1b), respectively, four resonances are observed for the random copolymer (Figure 1c). By comparison with the carbonyl chemical shift in PCL and PγBrCL,the resonances at 173.5 ppm and at 172.4 ppm are typical of the homodiads C-C and B-B, respectively. In the C-B heterodiad, the bromine atom triggers an upfield resonance at 173.2 ppm. In the case of the second heterodiad B-C, a downfield shift occurs at 172.6 ppm.

When the copolyester composition is changed, only the relative intensity of the four resonances changes, in contrast to the number and position of the carbonyl resonances which remain unmodified. In binary copolymers (C and B comonomers), the average lengths L_C and L_B can be calculated from the integration of the diad signals in the case of quantitative ^{13}C-NMR, according equations (1) and (2) where I stands for the integral of the diad under consideration.[6]

$$L_C = I_{CC} / [(I_{CB} + I_{BC}) / 2] + 1 \qquad (1)$$
$$L_B = I_{BB} / [(I_{CB} + I_{BC}) / 2] + 1 \qquad (2)$$

Table 2 shows that the F_B values determined by ^{1}H-NMR and ^{13}C-NMR are in good agreement within the limits of the NMR experimental errors, which confirms the assignment of the ^{13}C-NMR spectra.

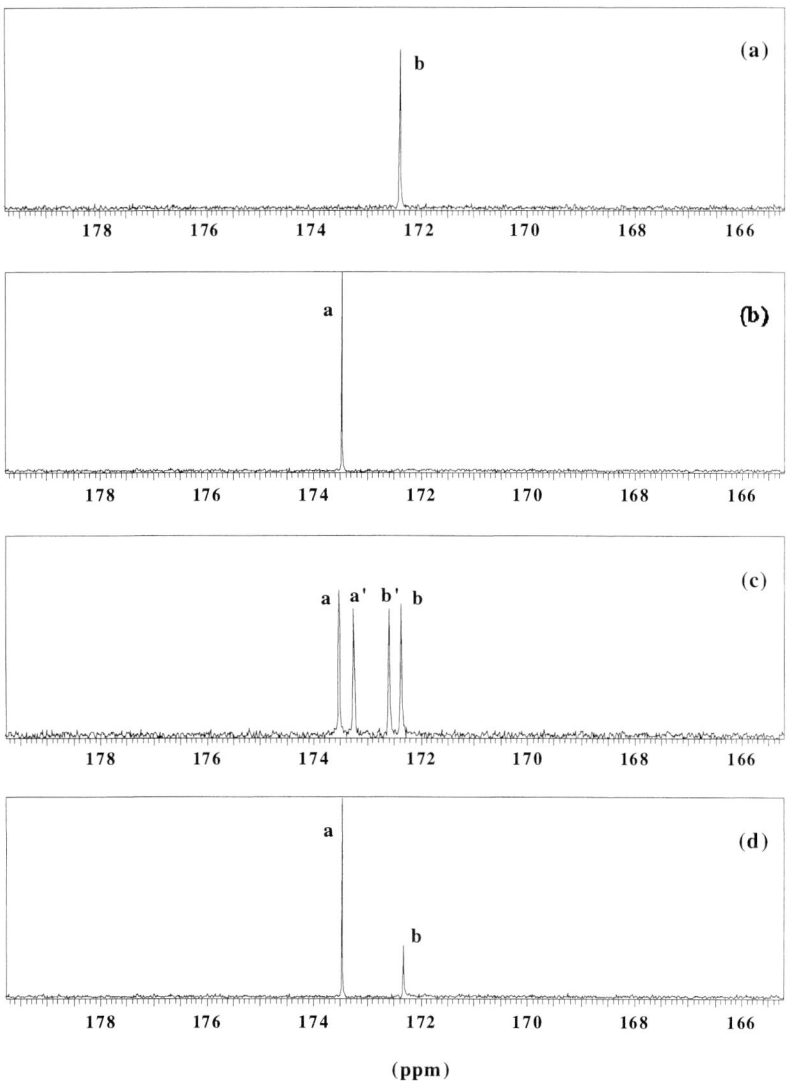

Figure 1 : ^{13}C-NMR spectra in the carbonyl region for PCL (a), PγBrCL (b), poly(γBrCL-co-εCL) (c) and poly(γBrCL-b-εCL) (d).

Table 2 : ^{1}H-NMR and ^{13}C-NMR data for poly(εCL-co-γBrCL).

Entry	F_B(^{1}H-NMR)	L_C	L_B	F_B(^{13}C-NMR)
1	0.09	10.77	1.08	0.09
2	0.28	3.18	1.58	0.33
3	0.49	2.15	2.09	0.49

$$F_B = L_B / (L_B + L_C) \quad (3)$$

A linear relationship is observed when the sequence length is plotted against $A = [\epsilon Cl]_0 / [\gamma BrCL]_0$ (Figure 2a) or against $A^{-1} = [\gamma BrCL]_0 / [\epsilon CL]_0$ (Figure 2b). The reactivity ratios, r_C and r_B, have been calculated from the slope of the straight line, according to the equations (4) and (5). $R_C=1.08$ and $r_B=1.12$.

$$L_C = (A \cdot r_1) + 1 \quad (4)$$
$$L_B = (r_2 / A) + 1 \quad (5)$$

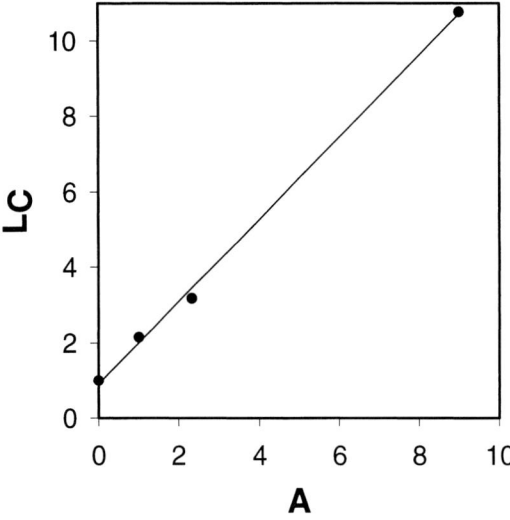

Figure 2a : Linear relationship of L_C versus A

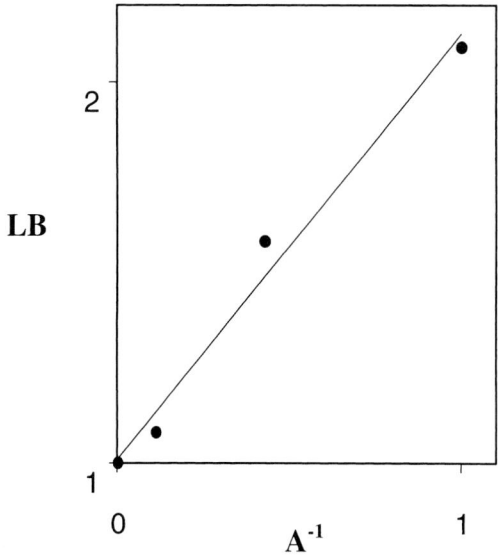

Figure 2b: Linear relationship of L_B versus A^{-1}

Although transesterification reactions can be responsible for errors in the L_C and L_B values, and thus in the r_C and r_B values, no reaction of this type is detected when the γBrCL polymerization is initiated by PCL macroinitiators in toluene at 0°C. Indeed, the blocky structure of the poly(ϵCL-b-γBrCL) copolymer is confirmed by the absence of the two heterodiad signals in the ^{13}C-NMR spectrum (Figure 1d). The narrow molecular weight distribution (M_w /M_n =1.2) of poly(ϵCL-co-γBrCL) and poly(ϵCL-b-γBrCL) is an additional evidence that no transesterification occurs during the sequential polymerization.

The thermal transitions of the poly(ϵCL-co-γBrCL) have been determined by DSC. The melting temperature of PCL decreases as the γBrCL content is increased until F_B of ca. 0.3 (Figure 3), consistently with the decreased average block length of the ϵCL units (L_C), reported in Table 2. At $F_B > 0.3$, L_C is indeed too small for crystallization to occur. The unique Tg which is observed increases with F_B as it could be anticipated from the higher Tg for PγBrCL compared to PCL (Tg=-60°C).

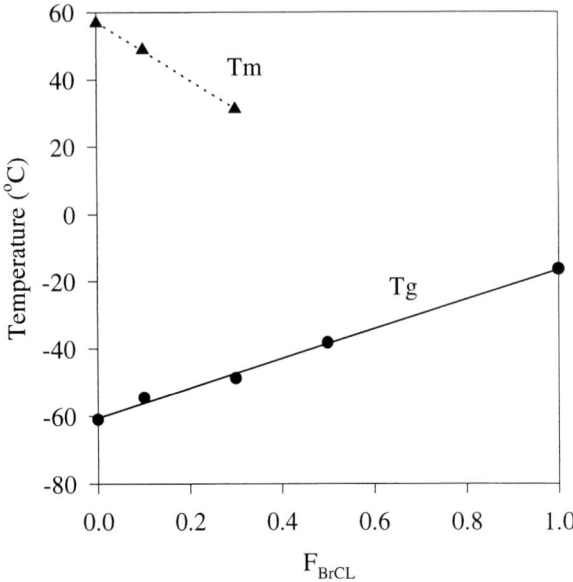

Figure 3 : Phase diagram for poly(CL-co-γBrCL)

Functionalization of PγBrCL

Reactivity of the γ–bromo substituent allows for new functional polyesters to be prepared. A polycationic polycaprolactone has been made available by reaction of pyridine and poly(εCL-co-γBrCL) (M_w/M_n = 1.2 ; $M_{n,\ H\text{-}NMR}^1$ = 16,500 ; F_B = 0.28) at 50°C (Scheme 5).[13] The quaternization is close to completeness (>90%) after 48 h. No olefinic proton is detected by ^{1}H-NMR, indicating that no elimination reaction has occurred. The average degree of polymerization calculated by ^{1}H-NMR before and after derivatization remains unchanged which is evidence that chains have not been degraded. This functionalization opens the way to the synthesis of hydrosoluble polyesters and to amphiphilic diblock copolymers. Further studies are in progress in this field.

The reaction of 1,8-diazabicyclo[5.4.0] undec-7-ene (DBU) with poly(εCL-co-γBrCL) (M_w/M_n = 1.2 ; $M_{n,\ H\text{-}NMR}^1$ = 17,000 ; F_B = 0.09) in toluene at 80°C leads to unsaturated polyester (Scheme 5).[13] The elimination reaction is not selective since a mixture of non-conjugated and conjugated olefinic units is formed. The molecular weight distribution remains narrow (M_w/M_n = 1.2), which is evidence of lack of degradation.

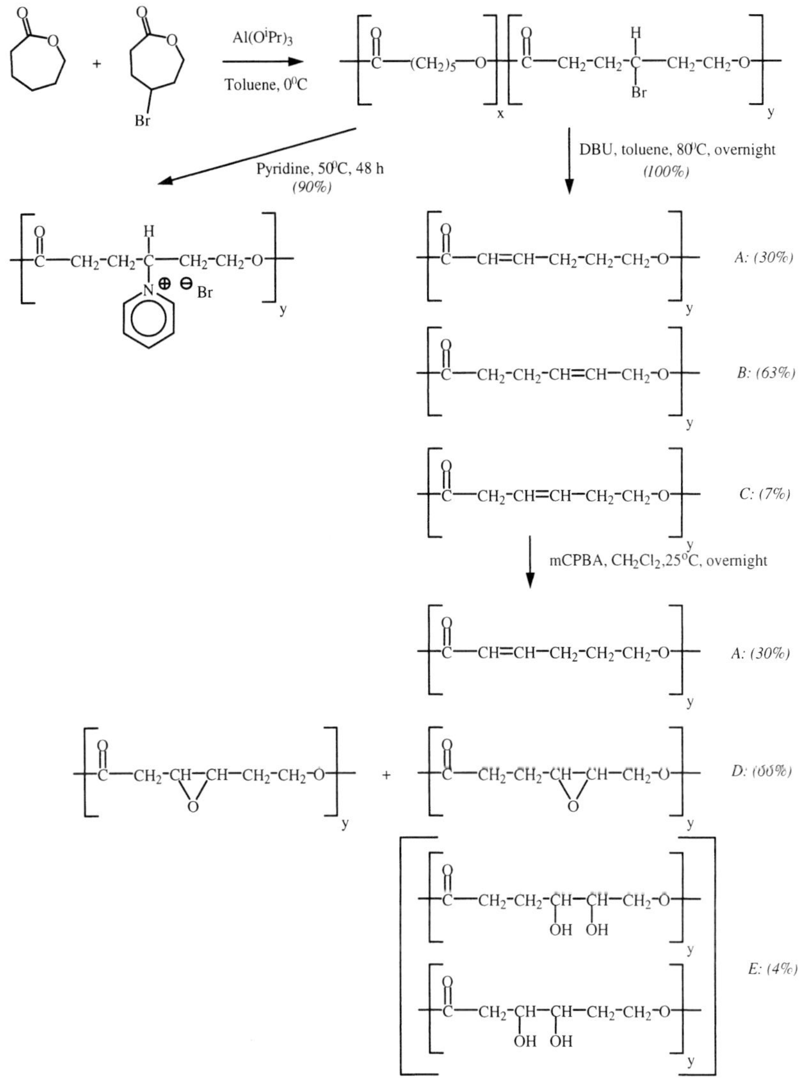

Scheme 5 : Quaternization and elimination of bromine in poly(εCL-co-γBrCL).

The quantitative epoxidation of the non conjugated double bonds has been carried out overnight at room temperature with m-chloroperbenzoic acid (mCPBA) in dichloromethane (scheme 5).[13] The conjugated double bonds remain untouched under these conditions. A low content (4%) of diol results from the acidic catalyzed hydrolysis of the epoxides. It must be noted that the epoxidation reaction does not lead to chain degradation and provides versatile intermediates for further functionalization of the polyester backbone. The non-epoxidized double bonds are also available to crosslinking reactions, which might be a way to prepare gels with tunable properties.

Unsaturated Lactones

Finally, an alternative pathway to unsaturated polyesters is proposed in Scheme 6, based on the reaction of γ–bromo–ε–caprolactone with DBU. Preliminary results show that the elimination reaction is not selective, leading to a mixture of three unsaturated lactones.

We are currently investigating a more selective synthesis of these new monomers. It must be noted that the double bond containing lactone could be polymerized by Ring Opening Metathesis Polymerization (ROMP) as a new route to unsaturated polyesters. This opportunity, including copolymerization with monomers typically polymerizable by ROMP, e.g. norbornene derivatives, will be studied in the near future.

ROP of the mixture of the unsaturated lactones has been initiated by $Al(OiPr)_3$ at room temperature (Scheme 6 ; Table 3). The first lactones mixture (Table 3, entry 1) has been polymerized almost quantitatively within 2 h. The second mixture (Table 3, entry 2), which contains 90% of lactone 1, has also been polymerized. Clearly, the polymerization is slower, which confirms that the conjugated double bonds in lactone 1 decrease the reactivity towards aluminum alkoxide. The experimental number-average molecular weights calculated by ^{1}H-NMR agree with the theoretical values ($M_{n,th}$) calculated from the initial monomer to initiator molar ratio. Furthermore, the polydispersity of the terpolymers is not exceedingly broad.

Table 3: ROP of mixtures of unsaturated lactones

Entry	Monomers 1 / 2 / 3	Time (h)	Yield (%)	$M_{n,th}$	$M_n,^1 {}_{H\text{-}NMR}$	$M_{n,SEC}$	M_w / M_n
1	26 / 24 / 50	2	94	11,000	11,000	14,000	1.35
2	90 / 5 / 5	14	60	4,000	3,500	3,000	1.4

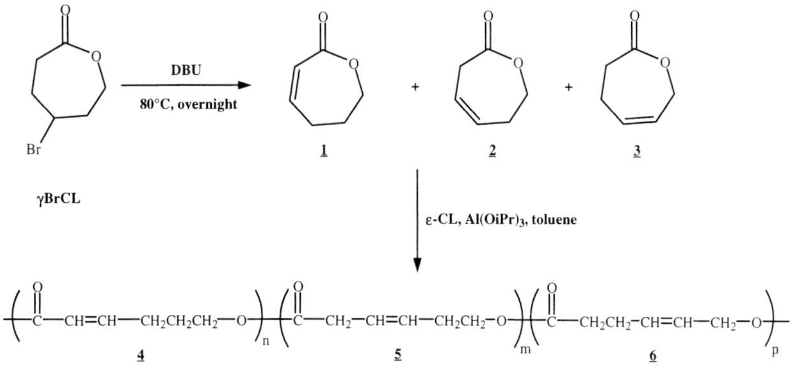

Scheme 6 : Elimination of bromine from γBrCL, and ROP of the unsaturated lactones

Conclusions

The controlled polymerization of functional ε-caprolactones is a versatile route to polyesters containing alcohol, ketone, bromo, olefinic and epoxy functions. Although some progress is needed to improve the selectivity of some derivatization reactions, a new set of polymers and copolymers is now available, whose the most representative properties will be investigated in the near future.

Acknowledgment

The authors are indebted to the « Services Fédéraux des Affaires Scientifiques, Techniques et Culturelles » for general support to CERM in the frame of the « PAI 4-11 : Supramolecular Chemistry and Supramolecular Catalysis ».

References

1   M. Vert, J. Feijen, A. C. Albertsson, G. Scott, E. Chiellini *Biodegradable Polymers and Plastics, Royal Society* (1992).

2   D. Mecerreyes, Ph. Dubois, R. Jérôme *Advances in Polymer Science,* in press.

3   D. Tian, Ph. Dubois, R. Jérôme, *Macromol. Symp.* **130**, 217 (1998) and references cited.

4   D. Tian, Ph. Dubois, Ch. Grandfils, R. Jérôme, *Macromolecules* **30**, 406 (1997).

5   D. Tian, Ph. Dubois, R. Jérôme, *Macromolecules* **30**, 1947 (1997).

6   D. Tian, Ph. Dubois, R. Jérôme, *Macromolecules* **30**, 2575 (1997).

7   D. Tian, O. Halleux, Ph. Dubois, R. Jérôme *Macromolecules* **31**, 924 (1998).

8   M. Trollsas, J. L. Hedrick, D. Mecerreyes, R. Jérôme, Ph. Dubois, *J. Polym. Sci., Polym. Chem.* **36**, 3187 (1998).

9   M. Trollsas, J. L. Hedrick, D. Mecerreyes, Ph. Dubois, R. Jérôme, H. Ihre, A. Hult, *Macromolecules* **31**, 2756 (1998).

10   G. Pitt, Z. W. Gu, P. Ingram, R. W. Hendren *J. Polym. Sci.: Polym. Chem., Part A* **25**, 955 (1987).

11   F. Stassin, O. Halleux, Ph. Dubois, Ch. Detrembleur, Ph. Lecomte, R. Jérôme, *Macromolecular Symposia* , accepted for publication.

12   Ch. Detrembleur, M. Mazza, O. Halleux, Ph. Lecomte, D. Mecerreyes, J. L. Hedrick, R. Jérôme, *Macromolecules* , submitted for publication.

13   Ch. Detrembleur, M. Mazza, O. Halleux, Ph. Lecomte, D. Mecerreyes, J. L. Hedrick, R. Jérôme, *Macromolecules,* to be published.

On the Mechanism of Polymerization of Cyclic Esters Induced by Tin(II) Octoate

Stanislaw Penczek[*], Andrzej Duda, Adam Kowalski, Jan Libiszowski, Katarzyna Majerska, Tadeusz Biela

Department of Polymer Chemistry, Centre of Molecular and Macromolecular Studies, Polish Academy of Sciences, 90-363 Lodz, Sienkiewicza 112, Poland

SUMMARY: Mechanism of initiation and propagation in polymerization of ε-caprolactone and L,L-dilactide induced with tin(II) octoate ($Sn(Oct)_2$) and $Sn(Oct)_2$/*n*-butyl alcohol system is presented. Tin(II) alkoxide bond formation is required in reaction of $Sn(Oct)_2$ with hydroxyl group containing compound to form a true initiator. Then tin(II) alkoxide end group is an active centre in the further propagation.

Introduction

Polymerization of cyclic esters, mostly ε-caprolactone (CL) and L,L-dilactide (LA) is gaining increasing interest since the corresponding polymers are degradable and have become industrial reality[1-7]. Therefore monomers are easier accessible and numerous blocks, grafts, hyperbranched polymers, based on these monomers are being developed[8].

There are two classes of initiators the most often used in the ring opening polymerization of cyclic esters: metal alkoxides and metal carboxylates. Initiation with the first class and further chain propagation is relatively well understood; all of the alkoxide substituents at the metal atoms start the growing chains, the rate of initiation for several metal alkoxides is comparable with the rate of propagation[9]. Tetramer of the aluminium *tris*-isopropoxide is the only well documented exception[10-12].

The side reactions revealed till now are reversible chain transfers to macromolecules with chain scission (via transesterification): either unimolecular (back biting) or bimolecular (chain transfer with chain rupture - reshuffling)[13-18]. In polymerizations with covalent metal alkoxides as active centres back-biting is suppressed kinetically[13-15], whereas reshuffling is not depriving polymerization processes from their living character[16-18]. Thus, the number of chains started does not change throughout the whole polymerization process, and all of the macromolecules initiated retain their ability to grow.

Metal carboxylates, mostly tin(II) octoate (Sn(Oct)$_2$), and less often used zinc derivatives, are very versatile initiators[19-35]. Commercial products can be handled in the half open system (i.e. do not require high vacuum equipment) and are relatively easy to purify (at least down to ≈2 mol-% of proton containing impurities)[35] by simple distillation for the synthetic applications and for semiquantitative work. The mechanism of initiation and the details of the chain growth were however not studied carefully enough in the past and a number of explanations were proposed, found recently[34-36] to be not fully correct.

In our recent papers[34-38] we described a series of experiments that allowed description of the general mechanism of polymerization initiated with Sn(Oct)$_2$. In the present short review we summarize the most important findings of the already published papers and add some additional evidence supporting the proposed mechanism.

The presence of Sn atoms in the macromolecules

The major questions to be answered and being a source of a certain controversy in the past are:
- is the Sn atom located, and in which form, in the growing macromolecules ?
- is Sn(Oct)$_2$ initiating by "itself"; i.e. by a direct reaction with monomers or whether it requires a coinitiator ?

The first question was answered by studying MALDI-TOF mass spectra of poly(ε-caprolactone) (PCL) and poly(L-lactide) (PLA) prepared in the presence of Sn(Oct)$_2$ and C$_4$H$_9$OH (BuOH) or H$_2$O, used as coinitiators and/or transfer agents (the role of coinitiator is described below). In the case of the CL/Sn(Oct)$_2$/H$_2$O system the following major populations of macromolecules have been observed (schematically) [34,36]:

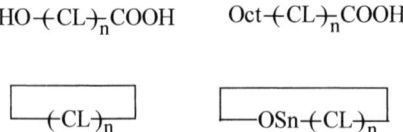

whereas for the CL/Sn(Oct)$_2$/BuOH system there was a larger number of various populations, shown below:

$$\text{OctSnO} \left(\text{CL} \right)_n \text{COOBu}$$

$$\text{HO} \left(\text{CL} \right)_n \text{COOBu} \qquad \text{Oct} \left(\text{CL} \right)_n \text{COOBu}$$

$$\text{HO} \left(\text{CL} \right)_n \text{COOH} \qquad \text{Oct} \left(\text{CL} \right)_n \text{COOH}$$

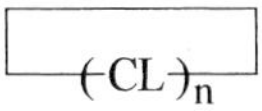

$$\left(\text{CL} \right)_n$$

Thus, for both systems macromolecules with tin(II) alkoxide units were observed :

$$...\text{-Sn-O-}(\text{CH}_2)_5\text{COO-}...$$

In the polymerization of LA with Sn(Oct)$_2$/BuOH system macromolecules fitted with OctSn-O-CH(CH$_3$)COO-... active end groups were also observed, similary to the polymerization of CL[38].

Thus, the first question - whether Sn atoms are on the chains has been answered.

Some elements of kinetics of polymerization with Sn(Oct)$_2$

Polymerization of CL or LA induced by Sn(Oct)$_2$ without a coinitiator purposely added is a very slow process (cf. kinetic plots in Fig. 1) and the rate clearly depends on the purity of Sn(Oct)$_2$ used. The better purity - the lower the rate of polymerization. We can safely assume (vide infra) that Sn(Oct)$_2$ alone does not initiate polymerization, at least at moderate temperatures. In a number of papers dealing with the mechanistic aspects of either CL or LA polymerization commercial Sn(Oct)$_2$ was used[23,32]. It contains, according to our ^{1}H NMR measurements, up to 30 mol% of compound(s) with active protons. We were able to purify Sn(Oct)$_2$ down to 0.9 mol-% (with respect to the octoate groups content) of these compounds, including presumably octanoic acid and water[35].

Polymerization of CL as well as LA induced by Sn(Oct)$_2$ in the presence of BuOH proceeds kinetically like a living process at least below 100°C in THF solvent and the dependence of M_n, of the resulting polyesteras a function of monomer conversion is a straight line as was described previously[35,37].

When BuOH (or e.g. $C_4H_9NH_2$ ($BuNH_2$)) are added to the polymerizing mixture, as we have shown for both CL[34,35] and LA[38], then the rates of polymerization increases, almost linearly at the lower $[BuOH]_0/[Sn(Oct)_2]_0$ ratio. However, when the ratio $[BuOH]_0/[Sn(Oct)_2]_0$ exceeds a certain value, then the rate becomes independent on $[BuOH]_0$. Similar observation was noted when the rate of polymerization was plotted as a function of $[Sn(Oct)_2]_0/[BuOH]_0$ ratio, i.e. when for a given set of kinetic measurements starting concentration of BuOH is kept constant in all experiments and starting concentration of $Sn(Oct)_2$ varies. This results is the most instructive, and meaning, that $Sn(Oct)_2$ as such is not providing active sites. The latter are formed in a reaction with BuOH that eventually gives an actual initiator.

Reaction of an alcohol with tin(II) carboxylate can be presented in the following way:

$$RC(O)OSnO(O)CR \ + \ R'OH \ \rightleftharpoons \ RC(O)OSnOR' \ + \ RC(O)OH \qquad (1a)$$

$$RC(O)OSnOR' \ + \ R'OH \ \rightleftharpoons \ R'OSnOR' \ + \ RC(O)OH \qquad (1b)$$

Whether both equilibria (1 (a) and 1 (b)) take place at the polymerization conditions or only the former - it is not clear at the present moment.

If this is indeed this exchange that provides tin(II) alkoxides as initiator, then the rate of polymerization with a system $Sn(O(O)CR)_2 + R'OH$ should be comparable to the rate of polymerization with a system $Sn(OR')_2 + RC(O)OH$ since:

$$R'OSnOR' \ + \ RC(O)OH \ \rightleftharpoons \ RC(O)OSnOR' \ + \ R'OH \qquad (2a)$$

$$RC(O)OSnOR' \ + \ RC(O)OH \ \rightleftharpoons \ RC(O)OSnO(O)CR \ + \ R'OH \qquad (2b)$$

Indeed, Schemes 1 and 2 are identical although approached from opposite sides and at a certain starting ratios of $Sn(O(O)CR)_2$ to HOR' (Scheme 1) and $Sn(OR')_2$ to RC(O)OH (Scheme 2) should give identical overall compositions. In Fig.1 we show that, for example for LA polymerizations, ratios: $[Sn(Oct)_2]_0/[BuOH]_0$ and $[Sn(OBu)_2]_0/[OctH]_0$ being close to 1:2 give almost identical rates of polymerization.

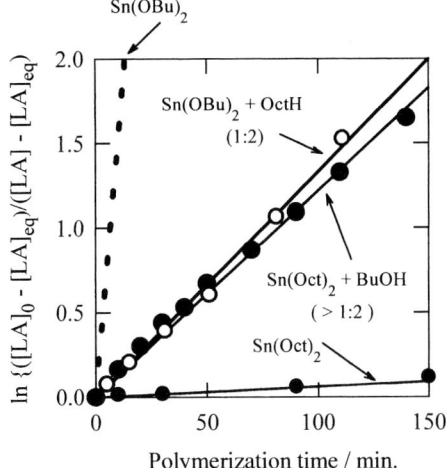

Fig.1: Kinetic convergence of Sn(OBu)₂/OctH and Sn(Oct)₂/BuOH initiating systems in the polymerization of LA. Polymerization conditions: $[Sn(Oct)_2]_0 = [Sn(OBu)_2]_0 = 0.05$ mol·L⁻¹, $[LA]_0 = 1.0$ mol·L⁻¹, THF solvent, 50°C (ref.[38]).

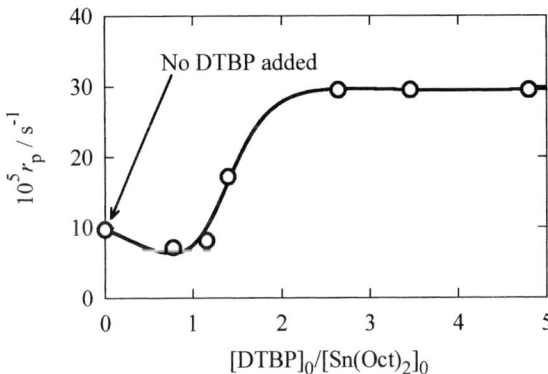

Fig.2: Infuence of 2,6-di(*tert*-butyl)pyridine (DTBP), used as a "proton trap", on the rate of polymerization of LA initiated with Sn(Oct)₂. Polymerization conditions: $[LA]_0 = 1.0$ mol·L⁻¹, $[Sn(Oct)_2]_0 = 0.05$ mol·L⁻¹, in THF solvent at 80°C (ref.[38]).

On the other hand, existence of these equilibria, providing a steady-state concentration of the actually growing macromolecules, should be sensitive to the addition or removal of one of the component. Thus, for Scheme 1, addition of RC(O)OH into the system should reduce the rate, whereas removal of RC(O)OH should increase the rate. Both phenomena were experimentally verified[35,38].

In Fig.2 the influence of a "proton trap" on the rate of polymerization is shown. After a sharp original decrease, there is an increase of rate until further addition of the hindered amine is no more influencing the rate. This result means that in Scheme1 equilibrium is shifted to the right hand side, because the acid is removed by complexing with the proton sponge. Whether both equilibria (1 (a) and 1 (b)) are involved - again it is not yet clear.

These are the most important kinetic results for the polymerization of CL and LA induced by $Sn(Oct)_2$ with coinitiators.

Mechanism of cyclic esters polymerization induced with $Sn(Oct)_2$

Kinetic and structural results (MALDI-TOF) taken together allowed to proposing the following mechanism of initiation, propagation, and formation of the end groups.

Initiation:

(a) preinitiation - formation of the true initiator:

$$RC(O)OSnO(O)CR \; + \; R'OH \quad \overset{fast}{\rightleftharpoons} \quad RC(O)OSnOR' \; + \; RC(O)OH \qquad (3)$$

(only the first equilibrium is shown from Scheme 1)

(b) first monomer addition:

$$RC(O)OSnOR' \; + \; M \quad \rightleftharpoons \quad RC(O)OSnO\text{-}m\text{-}R' \qquad (4)$$

(where M denotes the cyclic ester and m – the polyester repeating unit derived from M)

Propagation:

$$RC(O)OSnOR' \; + \; nM \quad \rightleftharpoons \quad RC(O)OSnO\text{-}(m)_n\text{-}R' \qquad (5)$$

Reversible chain transfer

(to the ROH which has been not used in initiation, before the chain transfer takes place):

$$RC(O)OSnO\text{-}(m)_n\text{-}R' \; + \; R'OH \quad \rightleftharpoons \quad RC(O)OSnOR' \; + \; HO\text{-}(m)_n\text{-}R' \qquad (6)$$

(starts a new chain)

Chain transfer with chain rupture (transesterification)

These processes are described in detail in our works devoted polymerization of CL and LA with metal alkoxides[13-18]. There is no difference when $Sn(Oct)_2$ is used, since polymerization proceeds eventually on similar active species.

Esterification of the hydroxyl end groups

Esterification of the hydroxyl end groups was documented by observation of the corresponding populations of PCL and PLA in MALDI-TOF[34,36,38]. The pertinent reactions, followed by propagation, read:

$$OctH + ROH \quad \xrightleftharpoons{\quad Sn(Oct)_2 \quad} \quad OctR + H_2O \tag{7a}$$

$$Sn(Oct)_2 + H_2O \quad \rightleftharpoons \quad OctSnOH + OctH \tag{7b}$$

$$OctSnOH + nM \quad \rightleftharpoons \quad OctSnO\text{-}(m)_n\text{-}H \tag{7b}$$

Assuming that eventually this process is irreversible (water is consumed) the proportion of the esterified end groups has a limit given by $[Sn(Oct)_2]_0$. In the technology of PLA the amount of $Sn(Oct)_2$ is low (tens of ppm) and the molar masses of the resulting polymer are close to 10^5. The starting concentration of monomer is equal to ≈ 8 mol·L^{-1} (polymerization in bulk). Thus, ≈ 1 % of macromolecules would have esterified end groups, provided, that the system is kept long enough in order to consume all of octanoic acid.

However, in some synthetic work, i.e. in preparation of block copolymers, starting from the long blocks, fitted with hydroxyl groups , one has to be aware that important part of these hydroxyl groups could be not accessible due to esterification. If, e.g. 1 mol-% of $Sn(Oct)_2$ is used, starting block of, e.g. α,ω-dihydroxy poly(ethylene oxide) has $M_n = 10^4$ and then $Sn(Oct)_2$ is introduced first, and kept for a long time before the monomer is added, then only 80 mol-% of these -OH groups is expected to be used in the polyester block formation.

End groups

Besides the octoate end groups, formed from one end of macromolecules, and discussed in the previous paragraph, there are other end groups, in several populations of macromolecules. Below, in Scheme 8 the mechanisms responsible for the formation of these populations are given:

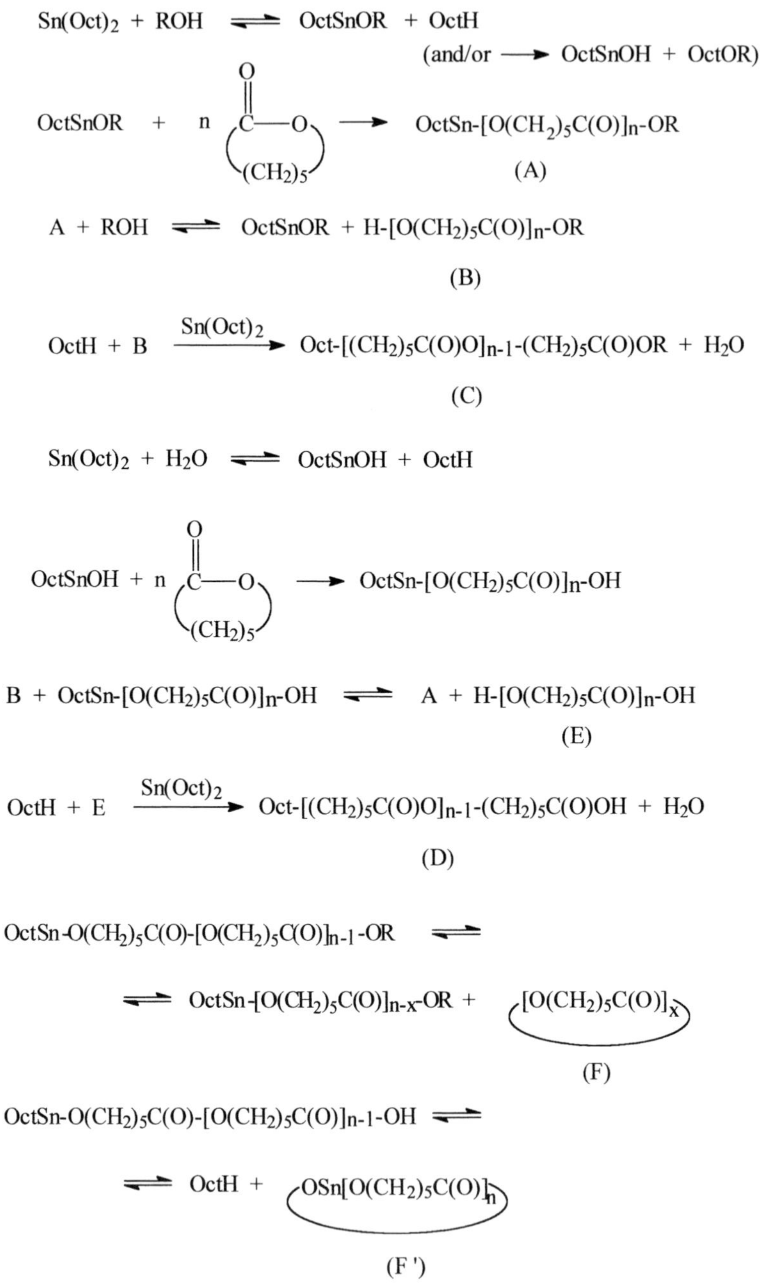

$$Sn(Oct)_2 + ROH \rightleftharpoons OctSnOR + OctH$$

$$(and/or \longrightarrow OctSnOH + OctOR)$$

$$OctSnOR + n \underset{(CH_2)_5}{\overset{O}{\overset{\|}{C}-O}} \longrightarrow OctSn-[O(CH_2)_5C(O)]_n\text{-}OR$$

(A)

$$A + ROH \rightleftharpoons OctSnOR + H\text{-}[O(CH_2)_5C(O)]_n\text{-}OR$$

(B)

$$OctH + B \xrightarrow{Sn(Oct)_2} Oct\text{-}[(CH_2)_5C(O)O]_{n\text{-}1}\text{-}(CH_2)_5C(O)OR + H_2O$$

(C)

$$Sn(Oct)_2 + H_2O \rightleftharpoons OctSnOH + OctH$$

$$OctSnOH + n \underset{(CH_2)_5}{\overset{O}{\overset{\|}{C}-O}} \longrightarrow OctSn\text{-}[O(CH_2)_5C(O)]_n\text{-}OH$$

(8)

$$B + OctSn\text{-}[O(CH_2)_5C(O)]_n\text{-}OH \rightleftharpoons A + H\text{-}[O(CH_2)_5C(O)]_n\text{-}OH$$

(E)

$$OctH + E \xrightarrow{Sn(Oct)_2} Oct\text{-}[(CH_2)_5C(O)O]_{n\text{-}1}\text{-}(CH_2)_5C(O)OH + H_2O$$

(D)

$$OctSn\text{-}O(CH_2)_5C(O)\text{-}[O(CH_2)_5C(O)]_{n\text{-}1}\text{-}OR \rightleftharpoons$$

$$\rightleftharpoons OctSn\text{-}[O(CH_2)_5C(O)]_{n\text{-}x}\text{-}OR + \overbrace{[O(CH_2)_5C(O)]_x}$$

(F)

$$OctSn\text{-}O(CH_2)_5C(O)\text{-}[O(CH_2)_5C(O)]_{n\text{-}1}\text{-}OH \rightleftharpoons$$

$$\rightleftharpoons OctH + \overbrace{OSn[O(CH_2)_5C(O)]_n}$$

(F ')

The kinetics and thermodynamics of the first equilibrium is still under study.

Conclusions

According to the kinetic data and direct observation of the macromolecules fitted with OctSn-O-polyester active end groups, polymerization of cyclic esters induced by $Sn(Oct)_2$ requires a preliminary formation of the tin(II) alkoxide bond, on which propagation proceeds. This course of reaction was either proposed or merely mentioned earlier in a number of papers[19,28,32] but clear cut evidence (like MALDI-TOF spectra provided by our work) was not available. Some recently published papers described another mechanisms shown in our recent paper to be unacceptable[38]. Thus, the major reaction, responsible for initiation is fast established equilibrium (first step):

$$RC(O)OSnO(O)CR \ + \ R'OH \ \overset{fast}{\rightleftharpoons} \ RC(O)OSnOR' \ + \ RC(O)OH \qquad (9)$$

which is shifted to the left hand side.

Acknowledgement: The financial support of the State Committee for Science (KBN), grant **3 T09B 105 11**, is gratefully acknowledged.

References

1. G. B. Kharash, F. Sanchez-Riera, D. K. Severson, in: *Plastics from Microbes,* D. P. Mobley (Ed.), Hanser Publishers, Munich, New York 1994, p. 93
2. M. H. Hartmann, in: *Biopolymers From Renewable Resources*, D. L. Kaplan (Ed.) Springer, Berlin 1998, p. 367
3. X. Zhang, U. P. Wyss, D. Pichora, M. F. A. Goosen, *J.Macromol.Sci.-Pure Appl.Chem.* **A30**, 933 (1993)
4. J. P. Benoit, C. Thies, in: *Microencapsulation. Methods and Industrial Applications,* S. Benita, (Ed.), Marcell Dekker, New York 1996, p. 133
5. R. G. Sinclair, *J.Macromol.Sci.-Pure Appl.Chem.* **A33**, 585 (1996)
6. J. Lunt, *Polymer Degrad.Stab.* **59**, 145 (1998)
7. M. Ajioka, K. Enomoto, K. Suzuki, A. Yamaguchi, *Bull.Chem.Soc.Jpn.* **68**, 2125 (1995)
8. D. Mecerreyes, R. Jerome, P. Dubois, *Adv.Polym.Sci.* **147**, 1 (1998)
9. A. Duda, S. Penczek, *Am.Chem.Soc., Symp.Ser.* 1999, in press.
10. A. Duda, S. Penczek, *Macromol.Rapid Commun.* **196**, 67 (1995)
11. A. Duda, S. Penczek, *Macromolecules* **28**, 5981 (1995)
12. A. Kowalski, A. Duda, S. Penczek, *Macromolecules* **31**, 2114 (1991)
13. A. Hofman, S. Slomkowski, S. Penczek, *Makromol.Chem., Rapid Commun.* **8**, 387 (1987)
14. S. Penczek, A. Duda, S. Slomkowski, *Makromol.Chem., Macromol.Symp.* **54/55**, 31 (1992)
15. S. Penczek, A. Duda, *Macromol.Symp.* **107**, 1 (1996)
16. J. Baran, A. Duda, A. Kowalski, R. Szymanski, S. Penczek, *Macromol.Symp.* **128**, 241

(1998)

17. J. Baran, A. Duda, A. Kowalski, R. Szymanski, S. Penczek, *Macromol.Rapid Commun.* **18**, 325 (1997).

18. S. Penczek, A. Duda, R. Szymanski, *Macromol.Symp.* **132**, 441 (1998)

19. J. W. Leenslag, A. J. Pennings, *Makromol.Chem.* **188**, 1809 (1987)

20. A. J. Nijenhuis, D. W. Grijpma, A. J. Pennings, *Macromolecules* **25**, 6419 (1992)

21. Y. J. Doi, P. J. Lemstra, A. J. Nijenhuis, H. A. M. van Aert, C. Bastiaansen, *Macromolecules* **28**, 2124 (1995)

22. F. Chabot, M. Vert, S. Chapelle, P. Granger, *Polymer* **24**, 53 (1983)

23. G. Schwach, J. Coudane, R. Engel, M. Vert, *J.Polym.Chem., Part A: Polym.Chem.* **35**, 3431 (1997)

24. K. Jamshidi, R. C. Eberhard, S.-H. Hyon, Y. Ikada, *Polym.Prepr. (Am.Chem.Soc., Div.Polym.Chem.)* **28(1)**, 236 (1987)

25. H. R. Kricheldorf, I. Kreiser-Saunders, C. Boettcher, *Polymer* **36**, 1253 (1995)

26. H. R. Kricheldorf, D.-O. Damrau, *Makromol.Chem.***198**, 1753 (1998);**198**, 1767 (1998)

27. J. Dahlman, G. Rafler, *Acta Polym.* **44**, 103 (1993)

28. X. Zhang, D. A. MacDonald, M. F. A. Goosen, K. B. McCauley, *J.Polym.Sci., PartA: Polymer. Chem.* **32**, 2965 (1994)

29. X. Zhang, U. P. Wyss, D. Pichora, M. F. A. Goosen, *Polym.Bull. (Berlin)* **27**, 623 (1992)

30. P. J. A. In't Veld, E. M. Velner, P. van de Witte, J. Hamhuis, P. J. Dijkstra, J. Feijen, *J.Polym.Sci., Part A: Polym.Chem.* **35**, 219 (1997)

31. A. Schindler, Y. M. Hibionada, C. G. Pitt, *J.Polym.Sci., PartA: Polym.Chem.* **20**, 319 (1982)

32. R. F. Storey, A. E. Taylor, *J.Macromol.Sci.-Pure Appl.Chem.* **A35**, 723 (1998)

33. A. Duda, S. Penczek, *Macromolecules* **23**, 1636 (1990)

34. A. Kowalski, J. Libiszowski, A. Duda, S. Penczek, *Polym.Prepr. (Am.Chem.Soc., Div.Polym.Chem.)* **39(2)**, 74 (1998)

35. A. Kowalski, A. Duda, S. Penczek, *Macromol.Rapid Commun.* **19**, 567 (1998)

36. A. Kowalski, A. Duda, S. Penczek, *Macromolecules* **33**, 689 (2000)

37. A. Duda, S. Penczek, A. Kowalski, J. Libiszowski, *Macromol.Symp.* (RO(M)P '99, Mons (Belgium) Volume), in press

38. A. Kowalski, K. Majerska, A. Duda, S. Penczek, in preparation

Macromol. Symp. **157**, *71–76 (2000)*

Ring-opening polymerization by various ionic processes

Hartwig Höcker[*]and Helmut Keul

Textilchemie und Makromolekulare Chemie, RWTH Aachen, Veltmanplatz 8,

D-52062 Aachen, Germany

SUMMARY: The ring-opening polymerization of cyclic carbonates, lactones, ester amides, urethanes and ureas as well as selected copolymerization reactions yield a variety of new polymers with well-defined architecture. The investigation of the mechanism of ring-opening polymerization reactions with anionic, insertion and cationic initiators shows a number of peculiarities beside activated chain-end and activated monomer mechanisms such as transfer and termination reactions, ring-expansion polymerization as well as the thermodynamic una bility of polymerization in special cases.

Introduction

For physical investigations as well as for several applications a well-defined architecture of polymers is desirable. Thus, living polymerization reactions have gained great importance and have been realised to a high degree for anionic, cationic, radical, and metal complex initiated polymerizations. Ring-opening polymerization may also follow at least a quasi-living path, in particular when back-biting and trans-reactions are absent. Besides, it offers the unique possibility to obtain block-copolymers by linking structures which usually are obtained by chain growth reactions with those which commonly are obtained by step growth reactions. This is realised when a macromolecular initiator is employed for the ring-opening polymerization of suitable monomers.

Ring-opening polymerization of cyclic carbonates

Six-membered cyclic carbonates are polymerised with variety of anionic initators the alcoholate group being the active species. Random co-polymers are obtained when different six-membered cyclic carbonates are employed. Co-polymerization with ε-caprolactone yields

        CCC 1022-1360/00/$ 17.50+.50/0

blockcopolymers with a tapered structure inbetween the two blocks dimethyltrimethylene-carbonate (DTC) being polymerised first. In the co-polymerziation with pivalolactone (PVL) block-copolymers are obtained since whenever PVL forms the terminal unit the active group is a carboxylate group which is unable to open the carbonate ring[1]. In the co-polymerization of DTC with L-lactide (LLA), polylactide (PLLA) is formed first. Whenever a DTC unit is added to the active end it attacks the carbonyl carbon of a LLA unit to form an LLA-DTC-LLA triade. With the progress of polymerization an increasing amount of mixed diades is formed[2].

The co-polymerization of DTC with ε-amino caprolactam (ECLam) follows a peculiar mechanism. While DTC is polymerised an ECLam anion is formed which attacks the carbonyl group of a carbonate unit to form an acylated ε-caprolactam endgroup which reacts in a ring-opening fashion with the alcoholate endgroup of another polymer chain to result in a polyester urethane unit. With $MgBu_2$ the reaction is fast and the polyester urethane is observed already after 5 min. With $Al(OsecBu)_3$ the reaction is slow and the ester amide is only observed after 5 h. It would be conceivable that an ECLam anion would attack the carbonyl carbon of an ester group as well as of a urethane group to result in a polyester amide or a polyester urea; urea groups, however, were not observed; amide groups are observed to a very small amount[3].

Ring-opening polymerization of cyclic urethanes

Trimethylene urethane is obtained from amino propanol and diphenyl carbonate with $Bu_2Sn(OCH_3)_2$ as a catalyst at 140 °C. After distillation and crystallisation the urethane is obtained in a yield of 61 %.

With anionic initiators and Et_2Zn and other insertion catalysts in the bulk a polymer with urea and carbonate units is observed resulting from head-to-head and til-to-til reactions. In a similar way, the polycondensation of an open chain dimeric hydroxy urethane $HO(CH_2)_3NH-CO-O(CH_2)_3NHCO-OC_6H_5$ are observed due to transurethanisation reactions. Cationic initiators, however, such as BF_3-ether, triflic acid, or triflate result in a uniform polyurethane in high yield with $M_n = 30.000$. The reaction follows first-order kinetics in the beginning of the reaction which is homogeneous. At higher conversions the melt becomes heterogeneous and aberrations from first-order kinetics occur. Correspondingly, the number average molecular weight of the polymer increases linearly with conversion as long as the reaction is homogeneous.

With triflate the initiation is achieved by methylation of the carbonyl oxygen resulting in an immonium cation which attacks the carbonyl oxygen of a monomer with O-alkyl-cleavage. The propagation follows an activated chain-end mechanism. The terminal cyclic immonium group may also react with other nucleophiles than the monomer, e.g., acetate or triphenyl phosphane to form the resepective endgroups.

In the course of the reaction beside chain growth chain transfer and termination occur when instead of the carbonyl oxygen the nitrogene of a monomer is attacked by the living end. Then the chain is terminated by formation of an alkylated cyclic urethane endgroup and a protonated monomer which after reaction with trimethylene urethane forms a carbamate group that is decarboxylated to form an amino group being immediately protonated. Thus the transfer step is followed by a termination reaction[4)5)].

Instead of low molecular weight initiators mono- and bifunctionally grown polytetrahydrofuran may be used as an initiator. When reacted with the cyclic urethane a protonated imminocarbonate endgroup is formed which after evaporation of excess tetrahydrofuran initiates the polymerization of the cyclic urethane. Thus AB and ABA blockcopolymers are obtained[6)].

In a very similar way the cyclic tetramethylene urethane is polymerised. The polymerization, however, occurs already at 60 °C in the bulk which is well below of the ceiling temperature of tetrahydrofuran. Thus this monomer is even more suitable for the formation of the respective blockcopolymers.

In the contrast, the cyclic dimethyltrimethylene urethane is not polymerised in a ring-opening fashion. The respective polymer rather is readily depolymerised to form the cyclic oligomers and eventually the cyclic monomer. On the other hand, the polymer is not observed by polycondensation of the monomeric hydroxy urethane $HO-CH_2-C(CH_3)_2-CH_2-NH-CO-O-$

C_6H_5 which yields exclusively the cyclic monomer but by polycondensation of the dimeric hydroxy urethane $H[O-CH_2-C(CH_3)_2-CH_2-NH-CO-O]_2C_6H_5$. It should be mentioned however, that the copolycondensation of the linear mono- and diurethane is possible yielding not only the even but also the odd members of the homologous series[7].

Cyclic diurethanes consisting of a tetramethylene diamine unit and a neopentylglycolbischloroformiate unit are readily polymerised with $Ti(OiPr)_4$ as well as with $Bu_2Sn(OMe)_2$ yielding polymers with an isopropyl endgroup and a methyl endgroup, respectively[8].

Ring-opening polymerization of cyclic ureas

Five- and seven-membered cyclic ureas and their polymerization with NaH in the melt were reported already in the fifties. The seven-membered cycle is readily polymerised in the melt at 210 °C as well as in DMF-solution with sBuLi as initiator at 140 °C. Dimethylene urea is polymerised with NaH at 140 °C in the melt following an activated monomer mechanism.

Tetramethylene urea is as well co-polymerised with DTC and $MgBu_2$ as an initiator both in the melt at 140 °C or in N,N'-dimethyl propylene urea (DMPU) at 120 °C. With increasing monomer concentration the molecular weight increases. From the ^{1}H-NMR spectrum of the co-polymers it is seen that DTC is polymerised with slight preference and the ^{13}C-NMR spectrum gives a clear indication of the tetrades formed. Thus, while DTC diades are decreasing DTC/urea diades are increasing and urethane diades are increasing to a low level. A very similar picture is obtained when living polyDTC is reacted with tetramethylene urea. This provokes a mechanism which is similar to that discussed in the co-polymerization of DTC and ECLam: a carbonyl carbon of the carbonate group is attacked by the cyclic urea-anion; the acylated urea endgroup then reacts with an alcoholate endgroup in ring-opening fashion to form a urethane group[9].

Surprisingly, tetramethylene urea is also copolymerised with ethylene carbonate to form a polyurethane which is clearly demonstrated by the ^{1}H-NMR spectrum. The cyclic tetramethylene urea anion reacts with dimethylene carbonate to form the acylated urea which then yields the alternating copolymer[10].

Ring-opening polymerization of cyclic ester amides

The most well known cyclic ester amides are cyclodepsipeptides which are obtained by reaction of α-hydroxy carboxylic esters with sodium salts of α-amino acids and subsequent cyclisation reaction[11]. The polymerization generally is achieved with $Sn(oct)_2$ or $Sn(acca)_2$ in the melt[12]. Beside cyclic depsipeptides dimeric cyclic ester amides consisting of two α-amino acids and two β-hydroxy acids in an alternating fashion are of some interest because of their antibiotic character (serratomolide). They are obtained from β-hydroxy acyl piperazines or just by cyclodimerisation reactions of the respective hydroxy acids or amino acid chlorides.

The chemistry ressembles that of an amide bond which when activated undergoes an N→O shift under ring enlargement when steric conditions are fulfilled. The fourteen-membered cyclic ester amides, however, do not undergo ring-opening polymerization with both $Ti(OiPr)_4$ and $Bu_2Sn(OMe)_2$ because of the high stability of the cyclic (mp 230-260 °C), the rigid and strainfree ring conformation stabilized by trans annular H-bonds.

In the contrast to the fourteen-membered dimeric cyclic ester amide the eleven-membered ester amide consisting of a β-hydroxy carboxylic acid and an ϵ-amino carboxylic acid was considered to be a suitable monomer. It is synthesized from an acroylated or a β-bromo or a β-benzyloxy acylated ϵ-caprolactam via ring expansion reaction[13]. With $Bu_2Sn(OMe)_2$ in DMF solution (120 h, 100 °C) ring-opening polymerization with 100 % conversion is achieved. With increasing monomer/initiator ratio the molecular weight increases. GPC clearly shows the presence of two types of oligomers indicating two competing initiation mechanisms. No cyclic oligomers are found. The molecular weight increases linear with conversion.

NMR indicates a fully alternating structure and two kinds of endgroups, i.e., a methyl ester endgroup and an acylated monomer endgroup in both cases the second endgroup being an alcohol endgroup. This corroborates the presence of two competing initiation mechanisms, i.e. the insertion of a catalyst into an acyl-O-bond and an activated monomer mechanism.

In the contrast to N-(3-hydroxypropionyl)-ε-amino caproic acid lactone the polymerization of N(3-hydroxypivaloyl)-ε-amino caproic acid lactone with naphthalene potassium, Bu_2Mg, $Al(OsecBu)_3$ and others as initiators follows exclusively a ring-expansion mechanism. After deprotonation of the monomer an equilibrium is achieved between the salt of the lactam and the salt of the cyclol. Then the anion performs a nucleophilic ring-opening of the monomer resulting - after protonation - in the key intermediate which intramolecularly forms the cyclol and after isomerisation the cyclic dilactame (cyclic dimeric ester amide). The molecular weight/conversion relation depends very much on the type of initiator used. The ^{13}C-NMR spectrum clearly shows the alternating structure of the polymer and GPC proves the cyclic nature of the oligomers as compared with those obtained via polycondensation of respective monomers which result in linear oligomers and polymers[14].

Conclusion

The ring-opening polymerisation of different monomers is characterised by a broad variety of unusual reactions opening a wide field in new polymers and in particular copolymers with well-defined architecture.

References

1. H. Höcker, H. Keul, *The Polymeric Materials Encyclopedia,* 1647-1654 (1996)
2. P. Schmidt, H. Keul, H. Höcker, *Macromolecules* **29**, 3674-3680 (1996)
3. B. Wurm, H. Keul, H. Höcker, *Macromolecules* **25**, 2977-2984 (1992)
4. S. Neffgen, H. Keul, H. Höcker, *Macromol. Rapid Commun.* **17**, 373-382 (1996)
5. S. Neffgen, H. Keul, H. Höcker, *Macromolecules* **30**, 1289-1297 (1997)
6. S. Neffgen, H. Keul, H. Höcker, *Macromol. Rapid Commun.* **20**, 194-199 (1999)
7. S. Neffgen, H. Keul, H. Höcker, *Macomol. Chem. Phys.* **199**, 197-206 (1998)
8. F. Schmitz, H. Keul, H. Höcker, *Macromol. Chem. Phys.* **199**, 39-48 (1998)
9. F. Schmitz, H. Keul, H. Höcker, *Polymer* **39**, 3179-3186 (1998)
10. F. Schmitz, H. Keul, H. Höcker, *Macromol. Rapid Commun.* **18**, 699-706 (1997)
11. V. Jörres, H. Keul, H. Höcker, *Macromol. Chem. Phys.* **199**, 825-833 (1998)
12. V. Jörres, H. Keul, H. Höcker, *Macromol. Chem. Phys.* **199**, 835-843 (1998)
13. B. Robertz, H. Keul, H. Höcker, *Macromol. Chem. Phys.* **200**, 1034-1040 (1999)
14. B. Robertz, H. Keul, H. Höcker, *Macromol. Chem. Phys.* **200**, 1041-1046 (1999)

Macromol. Symp. 157, 77–92 (2000)   77

THE INDANYL CATION: A REAL TIME OBSERVATION

M. Givehchi, A. Polton, M. Tardi, M. Moreau, P. Sigwalt and J. -P. Vairon

Laboratoire de Chimie Macromoléculaire - UMR 7610

Université Pierre et Marie Curie, 4, Place Jussieu, Paris, cedex 05, France

Abstract

A stopped-flow investigation by U. V. spectroscopy has been carried out using various reactions which yield the indanyl cation: polymerization of indene by trifluoromethane sulfonic acid (TfOH), ionization of 1-chloroindane by antimony pentafluoride and protonation of a dimer of indene (2-α-indanyl indene) by TfOH, at variable temperature. The monomer and dimer cations present a main absorption at 318-325 nm and the polyindene cation at 330 nm. A side reaction yields a derived cation, which absorbs at 519 nm. The molar absorbance of the indanyl cation has been estimated (ε = 15 500 L.mol.$^{-1}$ cm^{-1}).

INTRODUCTION

Indene is a promising monomer for controlled polymerization because it has a low transfer constant to monomer and does not undergo indanic termination. The controlled polymerization of indene has been achieved using TiCl$_4$ as activator and cumyl chloride [1,2] or methyl cumyl ether [2,3] as initiators, in the last case without additive. This suggests that the indanyl cation is relatively stable and does not undergo isomerization or destruction.

Hence it was interesting to characterize the indanyl cation involved in this polymerization and to check its stability. The characterization of the active species in cationic polymerizations of alkyl and arylalkyl monomers is fraught with difficulties, due to the reactivity of the cations, which quickly isomerize or are destroyed. In the case of indene, polymerization is often very fast, in some cases completed in a few seconds, and what can be observed might not necessarily be the propagating cation, but derived species.

The two main methods of investigation are U.V.visible spectroscopy and nuclear magnetic resonance. However, the timescale of the polymerization is generally much shorter than the time of measurement, and these methods can only be used in favourable cases in which the carbocation is sufficiently stable [4]. One possibility is to use a very large excess of initiating Lewis acid with respect to the monomer in order to favour initiation with respect to

          CCC 1022-1360/00/$ 17.50+.50/0

propagation or to operate in solution in concentrated sulfuric acid or "super acids". Another possibility is to use non-polymerizable, or poorly reactive, compounds having a structure similar to that of the monomer, such as substituted indenes in the case of indene.

Olah and coll. have used "super acids" to protonate a wide range of compounds[5]. For the indene series, most references in the literature refer to substituted indanyl cations. Deno et al. have investigated the protonation of 3-methylindene at room temperature by ^{1}H NMR [6] and by UV spectroscopy [7] and have reported an unexpected λ_{max} at 401nm ($\varepsilon = 1148$ L.mol.$^{-1}$ cm^{-1}). They also reported in the same work the chemical shifts of the 1,3,3-trimethylindanyl cation resulting from the cyclization of 1-methyl-4-phenyl-1,3-pentadiene in 96% SO_4H_2 .

A wide range of substituted indanyl cations has been investigated in our laboratory by UV spectroscopy and NMR, under high purity conditions[8,9]. The main results concerning the former technique are summarized in table 1. It appears that most 1-monosubstituted indanyl cations have absorbances in the range of 312-322 nm, with exceptions for the 1,4,5,6,7-pentamethylindanyl cation ($\lambda_{max} = 348$ nm), and for the 1-phenylindanyl cation which absorbs at a still higher wavelength ($\lambda_{max} = 412-418$ nm). All these cations have molar extinction coefficients of about 30000 L.mol.$^{-1}$ cm^{-1}. The cations derived from 3-phenylindene (1-phenylindanyl cation) [10,11] and from 3-isopropyl indene (1-isopropyl cation)[12] have also been investigated using ^{13}C and ^{1}H NMR respectively.

The unsubstituted indanyl cation appears to be more elusive than its substituted counterparts, and the references are rather scarce. Most attempts at measuring the absorption maximum of this cation did not yield clear-cut results. The first mention of the spectroscopy of an indanyl cation was made by Olah and coll. [13], which reported that protonation of indene by sulfuric acid at room temperature yields a species having two main absorptions at 301 nm ($\varepsilon = 14100$ L.mol.$^{-1}$ cm^{-1}) and 401 nm ($\varepsilon = 2200$ L.mol.$^{-1}$cm^{-1}). Ledwith and col. [14] using as an initiator tropylium hexachloroantimonate for the polymerization of indene observed a transient maximum at 435-440 nm which was ascribed to a "substituted indane like carbonium ion", while a second maximum at 525 nm was supposed to be a side product involving a tropylium cation attached to the polyindene chain. In an U. V. investigation of the polymerization of indene, Prosser et al. [15] have observed a maximum at 318 nm in sulfuric acid and a "transient species" at 340 nm. Furthermore, they concluded that another maximum at 515 nm corresponds to a cation resulting from hydride abstraction from the penultimate indene unit of an unsaturated end group.

Solutions of styrene or indene in hydrocarbons in the presence of zeolites yield carbocations. For instance, styrene dimerizes and yields a 3-methyl-1-phenylindanyl cation [16]. When an activated zeolite is added to a hexane solution of indene, a strong UV absorption appears at 450-550 nm, which remains stable at room temperature[17]. The authors have proposed a reaction scheme involving mainly the formation of dimers of indene.

Up to now, there is still no certainty concerning the absorption wavelength of the unsubstituted indanyl cation, the values reported varying from 301 to 340 nm.

The stopped-flow technique allows fast observation of the formation of the cations from the very beginning of the reaction, and to monitor their possible subsequent isomerizations. It is the most efficient mean of investigation in the case of fast polymerizations, such as that of indene. The first report concerning the use of a stopped-flow technique in cationic polymerization is from Pepper and coll [18] for the polymerization of styrene. This technique has been implemented to investigate the cations of styrene[19,20], pmethoxystyrene[21] and 1,1-diphenylethylene [22]. In the case of indene, Kunitake et al[23] mention a maximum at 404 nm (at 30°C) for the indanyl cation in a stopped-flow experiment. A stopped-flow apparatus allowing to operate at low temperature and under vacuum, under high purity conditions, has been built in our laboratory[24]. It has been used to investigate the polymerization of styrene[25], as well as the reactions of the dimers of styrene[26,27]. In the present paper, this technique has been put to use to investigate the active species formed in various reactions involving the indanyl cation: the polymerization of indene by trifluoromethanesulfonic acid (TfOH), the ionization of 1-chloroindane by antimony pentafluoride and the protonation of the dimer of indene (2-α-indanyl indene) by TfOH. Triflic acid is transparent above 250 nm and can be used in relatively large concentrations, which favours protonation.

Monomer	Conditions	T (°C)	λ_{max} (nm)	ε $(M^{-1}.cm^{-1})$
3-methyl indene	in H_2SO_4	20°C	312	27 000
3-methyl indene	$TiCl_4$ sol. in CH_2Cl_2	-70°C	318	30 000
3-isopropyl indene	in H_2SO_4	20°C	322	29 000
3-isopropyl indene	CF_3SO_3H sol. in CH_2Cl_2	-70°C	318	29 000
3-isopropyl indene	$TiCl_4$ sol. in CH_2Cl_2	-70°C	322	30 800
3-phenyl indene	H_2SO_4 sol. in CCl_4	20°C	412	32 000
3-phenyl indene	CF_3SO_3H sol. in CH_2Cl_2	-70°C	414	32 000
3-phenyl indene	$TiCl_4$ sol. in CH_2Cl_2	-70°C	418	32 000
3,4,5,6,7-pentamethyl indene	$TiCl_4$ sol. in CH_2Cl_2	-70°C	348	32 000

Table 1 : Absorbance and molar absorbance of substituted indanyl cations (ref. 8-11)

RESULTS

1) Protonation of indene by triflic acid

The polymerization of indene initiated either by Lewis acids or triflic acid is fast and completed within seconds or a few minutes. Consequently, protonation, dimerization and polymerization may take place simultaneously and the following set of reactions must be considered (scheme 1):

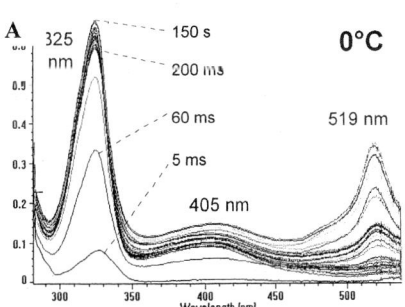

Scheme 1

Influence of temperature.

At 0°C, protonation of indene by TfOH ([TfOH] = $2 \; 10^{-3}$ mol.L^{-1}; [Indene] = $2 \; 10^{-3}$ mol.L^{-1}, R = [TfOH] / [Indene] = 1) yields in 5 milliseconds an absorbance at 330 nm which rapidly shifts to 325 nm (Figure 1) and reaches a maximum in 1 s (Figure 2). Two secondary absorbances appear at 405 nm and 519 nm. The maxima at 325 and 405 nm remain constant over 150 s, and that at 519 nm slightly increases with time. At -65°C, a single maximum at 330 nm is observed, with a low absorbance (Figure 3), which remains unchanged over 600 s.

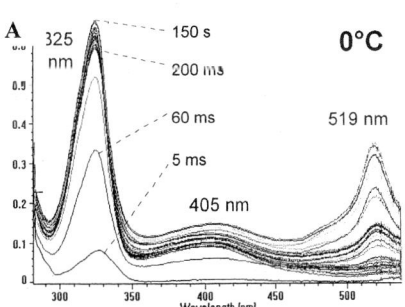
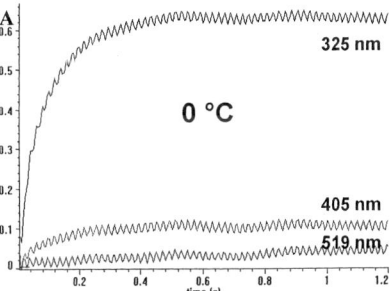

Figure 1 : Protonation of indene by triflic acid at 0°C. UV spectra of the solution. [Indene] = $2 \; 10^{-3}$ mol.L^{-1} ; [TfOH] = $2 \; 10^{-3}$ mol.L^{-1} ; (R=1); solvent: CH$_2$Cl$_2$

Figure 2 : Protonation of indene by triflic acid at 0°C. Evolution of the maxima with reaction time. [Indene] = $2 \; 10^{-3}$ mol.L^{-1}; [TfOH] = $2 \; 10^{-3}$ mol.L^{-1} ; (R=1); CH$_2$Cl$_2$

At all these temperatures, the initial absorption (10 ms) is at 330 nm. It remains stable over 600 s at -65°C, while at higher temperatures the position of the final absorption after 600 s progressively shifts towards lower wavelengths (Figure 4). The maximum absorptions of the three bands progressively decrease as the reaction temperature decreases from 0°C to -65°C (Figure 5).

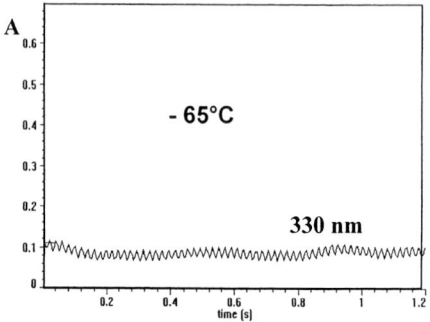

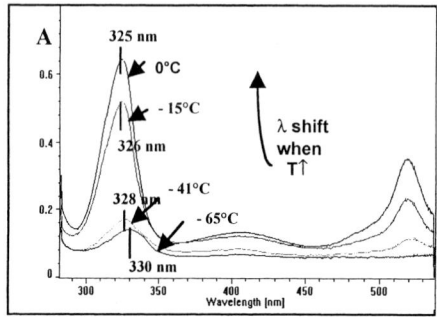

Figure 3 : Protonation of indene by triflic acid at –65°C. Evolution of the maximum with reaction time. (R=1) [Indene] = $2 \cdot 10^{-3}$ mol.L^{-1}; [TfOH]=2 10^{-3} mol.L^{-1}

Figure 4 : Protonation of indene by triflic acid at variable temperature. UV spectra after 600 s. [Indene] = $2 \cdot 10^{-3}$ mol.L^{-1}; [TfOH] = $2 \cdot 10^{-3}$ mol.L^{-1}; (R=1) ; CH$_2$Cl$_2$

Influence of initiator to monomer ratio

When the ratio of TfOH to indene is increased at -65°C ([TfOH] = $4 \cdot 10^{-3}$ M; [Indene] = $2 \cdot 10^{-3}$ M; R = 2), a single stable maximum at 325 nm is obtained (A = 0.1 instead of 0.06 for R = 1). As expected, an increase in R increases ionization. At 0°C, the maximum at 325 nm is much higher than at -65°C, and is reached in 0.3s (A = 0.85; Figure 6). Even at 0°C, the maximum absorbance at 325 nm is relatively stable (100 s) but slowly decreases over 600 s. At this temperature, another absorption band appears at 519 nm and continuously increases up to 1.0 in 600 s (Figure 7).

In order to check the reaction products formed during the reaction, a polymerization of indene has been carried out at 0°C ([TfOH] = $3.2 \cdot 10^{-3}$ mol.L^{-1} ; [Indene] = $2.1 \cdot 10^{-3}$ mol.L^{-1} ; R = 1.5) for 2 minutes and quenched with a methanol / pyridine mixture. After evaporation of the solvent, the reaction products were analyzed by size exclusion chromatography. The

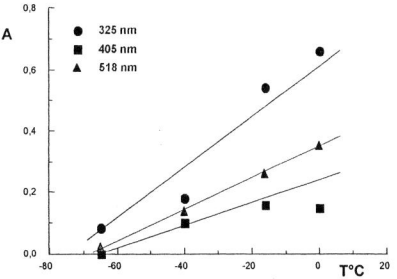

Figure 5 : Protonation of indene by triflic acid. Variation of the absorbance with temperature. [Indene] = $2\ 10^{-3}$ mol.L^{-1}; [TfOH] = $2\ 10^{-3}$ mol.L^{-1} solvent: CH_2Cl_2

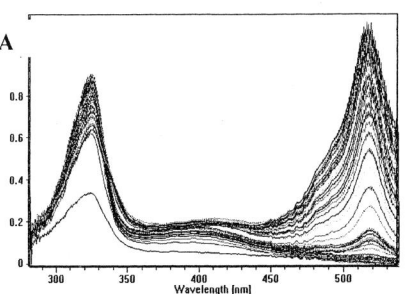

Figure 6 : Protonation of indene by triflic acid at 0°C. UV spectra of the solution. [Indene] = $2\ 10^{-3}$ mol.L^{-1} ; [TfOH] = $4\ 10^{-3}$ mol.L^{-1} ; (R=2); solvent: CH_2Cl_2

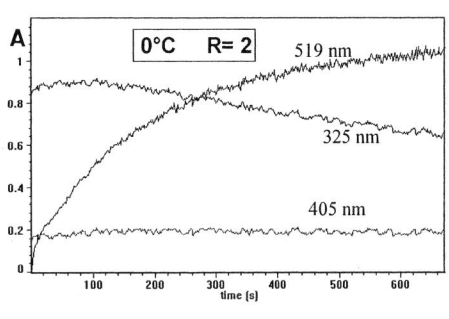

Figure 7 : Protonation of indene by triflic acid at 0°C. Evolution of the maxima with reaction time. (R=2) [Indene]= $2\ 10^{-3}$ mol.L^{-1}; [TfOH] = $4\ 10^{-3}$ mol.L^{-1}; CH_2Cl_2

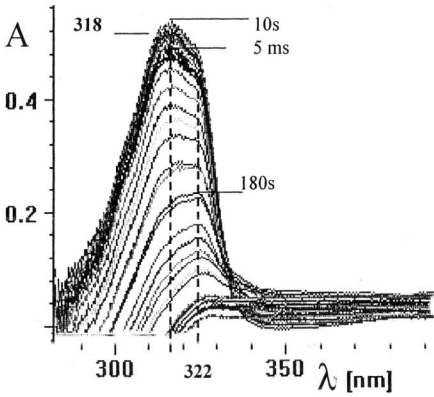

Figure 8 : Ionization of 1-chloroindane by SbF$_5$ at –68°C. UV spectra of the solution. [1-Chloroindane] = $1.4\ 10^{-4}$ mol.L^{-1}; [SbF$_5$] = 10^{-3} mol.L^{-1} (R= 7.1); CH_2Cl_2

chromatogram showed essentially oligomers of indene, but no indene monomer, which would have resulted from the monomeric cation. No trace of indene was found as well in the vapour phase analysis of the evaporate. This suggests that the maximum absorption at 325 nm, which appears at 0°C, does not correspond to the monomeric indanyl cation, but to dimeric or oligomeric species.

2) Ionization of 1-chloroindane by antimony pentafluoride.

Reaction of a Lewis acid such as antimony pentafluoride with 1-chloroindane should produce the indanyl cation (scheme 2):

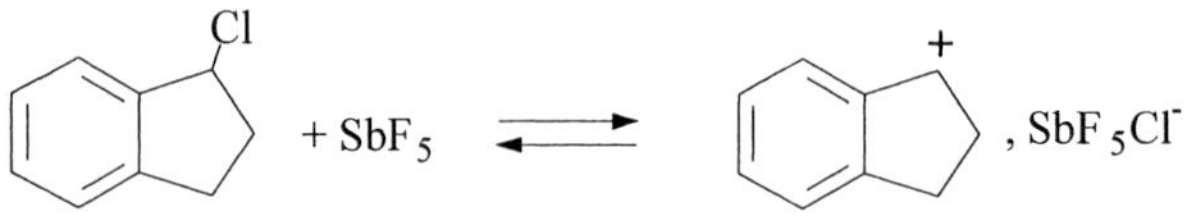

Scheme 2

When SbF_5 (1.5 10^{-3} mol.L^{-1}) is reacted with 1-chloroindane (1.4 10^{-4} mol.L^{-1} ; R = 7.1) at -65°C, a large absorption band rapidly appears at 314-322 nm, which reaches a maximum in a few seconds with no other visible absorption up to 540 nm (Figure 8). Closer examination of figure 8 shows the presence of a maximum at 318 nm and of a shoulder at 322 nm. The global absorption decreases with time to half the initial value in 180 s and the main absorption shifts to 322 nm.

3) Protonation of 2-α-indanyl indene (dimer) by triflic acid

In the case of the protonation of indene, polymerization takes place concurrently with ionization, and the species observed is a mixture of oligomeric cations, particularly at high temperatures. On the other hand, protonation of the dimer of indene (2-α-indanylindene) by TfOH should yield the dimeric indanyl cation (scheme 3). 2-substituted indenes do not polymerize easily and we have found that the dimer of indene does not polymerize and that no tetramer (i. e. dimer of dimer) is formed at -60°C after 10 min, in the presence of equimolar concentrations of TfOH and dimer (2.5 10^{-2} mol.L^{-1}). In th e same conditions, but at 0°C, a

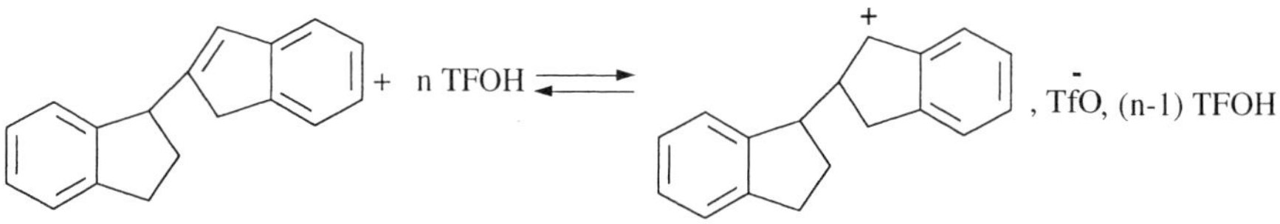

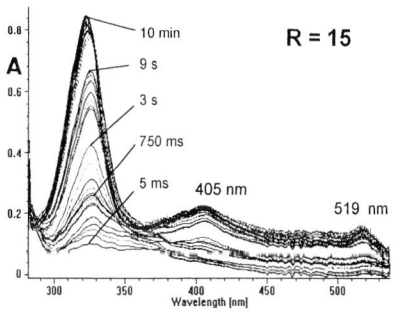

Scheme 3

small amount of oligomers is formed after 45 min, which is much longer than the time of observation (600 s in most cases) and than that of polymerizations of indene. Consequently, protonation of the dimer, which should yield a cation, which is the model of the growing polyindene chain, has been investigated.

At -66°C, for a high TfOH to dimer ratio (R = [TfOH] / [Dimer] = 15 ; [TfOH] = 4 10^{-3} mol.L^{-1}; [Dimer] = 2.6 10^{-4} mol.L^{-1}), a main absorbance at 324-327 nm is observed, which reaches a maximum in 50 s. Besides, the two absorptions at 405 and 519 nm observed in the case of the protonation of indene are present with a low absorbance (Figure 9). The three absorptions quickly reach a maximum value and remain unchanged over 600 s (Figure 10).

Figure 9 : Protonation of 2-indanyl indene by triflic acid at –66°C. UV spectra of the solution. (R=15) [Dimer] = 2.6 10^{-4} mol.L^{-1} [TfOH] = 4 10^{-3} mol.L^{-1} ; CH_2Cl_2

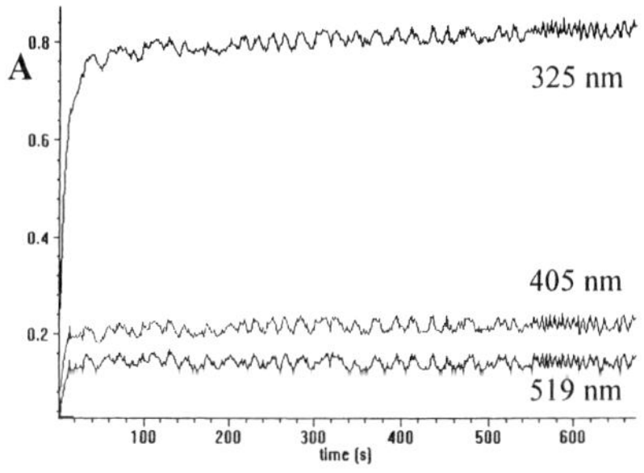

Figure 10 : Protonation of 2-indanyl indene by triflic acid at –66°C. Evolution of the maxima with time. [Dimer] = 2.6 10^{-4} mol.L^{-1} ; [TfOH] = 4 10^{-3} mol.L^{-1} (R=15)

A series of measurements has been carried out at -66°C with a constant dimer concentration (2.5 10^{-4} mol.L^{-1}) and increasing TfOH concentrations in order to favour protonation. For R ratios of 6 and 13, the three maxima at 325-328, 405 and 519 nm are observed. For higher values of this ratio (R = 26, 52, 100) there is only a main absorption at 325 nm, a small maximum at 405 nm and practically no peak at 518 nm. The incidence of the concentration of TfOH on the maximum absorbance of the three maxima at 325-328, 405 and 519 nm is shown in figure 11. The maximum at 325 nm increases as the concentration of triflic acid is increased and seems to reach a plateau (A_{325} = 0.7) for R = 52. For the two highest concentrations of TfOH, there is practically no change, which is in agreement with the hypothesis of an instantaneous quantitative protonation of the dimer. If one assumes that in this case the protonation equilibrium is strongly shifted towards protonation, the corresponding molar extinction coefficient would be 15500 L.mol^{-1}.cm^{-1}, which is a minimal value. The maximum at 405 nm also increases, but that at 518 nm decreases and is no longer visible for R = 52.

The rate of formation of the absorption at 325 nm depends on the concentration of TfOH. The maximum absorbance is reached in 150 s for R = 6, in 50 s for R = 13, and in less than two seconds for highest ratios. The value of the absorbance at 325 nm after 0.4 s has been measured. The ratio (A_{325} / reaction time) is proportional to the protonation rate. The corresponding plot of Ln (A_{325}/ 0.4) versus Ln ([TfOH]) shows that for the lower values of R (from 6 to 26), the reaction order with respect to TfOH is near one (1.16, see figure 12).

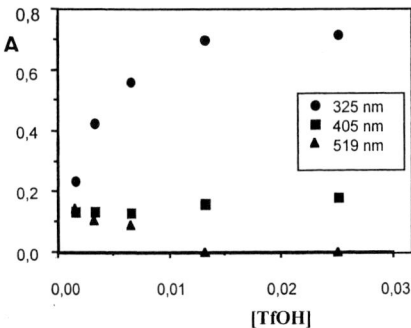
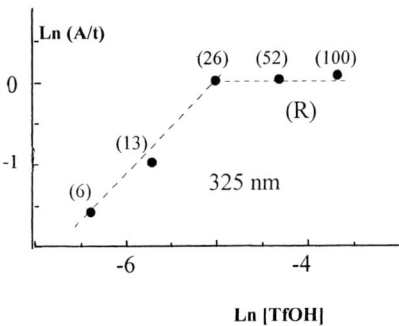

Figure 11 : Protonation of 2-indanyl indene by triflic acid at –66°C. Evolution of the maximum absorbances with TfOH concentration. [Dimer] = 2.6 10^{-4} mol.L^{-1}

Figure 12 : Protonation of 2-indanyl indene by triflic acid at –66°C. Determination of the reaction order with respect to TfOH. [Dimer] = 2.6 10^{-4} mol.L^{-1} ; solvent: CH$_2$Cl$_2$

The influence of temperature on the three maxima is shown in figure 13. Contrarily to the case of indene, the maximum at 325 nm increases at low temperature, as well as that at 405 nm, while that at 519 nm decreases.

DISCUSSION

In all these experiments, the main absorption maximum is observed in the 318-330 nm range, but its position depends on the experimental conditions. The slight differences in wavelength in the various experiments may be due to the presence of different types of indanyl cation (provided that they are not due to experimental conditions, such as the nature of the counterion).

Protonation of 2-α-indanyl indene, which should give a dimeric cation, yields a maximum at 325 nm, which is stable over 600 s.

Ionization of 1-chloroindane should yield the monomeric indanyl cation, which may correspond to the maximum at 318 nm. Besides, it produces more slowly a secondary species at about 322 nm (i. e. near the λ_{max} of the dimer) which could be the dimeric cation resulting from addition of monomer units formed by deprotonation of the indanyl cation (scheme 4):

$$1\text{-Cl Ind} + SbF_5 \;\rightleftharpoons\; M^+ \;\underset{M}{\overset{(-H^+)}{\rightleftharpoons}}\; M$$

$$(+M) \downarrow$$

$$D^+ \quad (+T^+ ...)$$

Scheme 4

This maximum is not stable and disappears in about 200 s, contrarily to what was observed in the case of the dimer. The decrease of this maximum may result from halogen exchange which yields non-reactivable 1-fluoroindane (scheme 5).

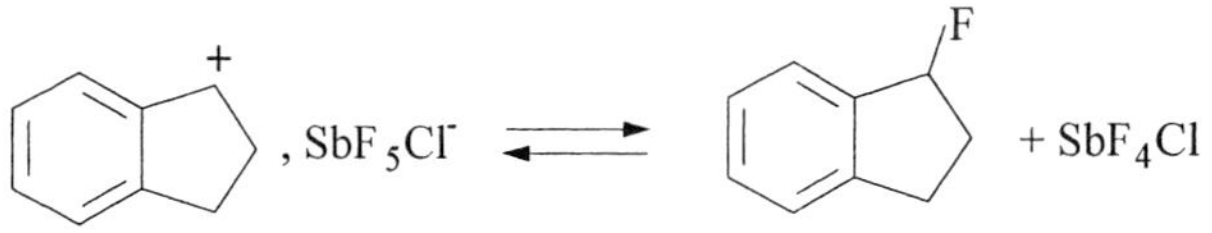

Scheme 5

The situation is more complicated in the case of the protonation of indene. In favourable conditions (low temperature, high [TfOH] / [Indene] ratios), a single stable low absorbance at 330 nm is observed. For higher temperatures, the initial maximum at 330 nm shifts to 325 nm with a much higher absorbance. A possible explanation would be that at -65°C polymerization quickly takes place, consuming a large amount of monomer and yielding few stable polymeric cations (low absorbance) and that there is no depolymerization at this temperature, while at 0°C depolymerization of indene is not negligible and eventually leads to oligomeric cations and to indene which would be cationated. The comparison of the influence of temperature on the maximum at 325-330 nm in the protonation of indene and of the dimer agrees with this hypothesis. In the case of the protonation of indene, the absorbance decreases at low temperature (Figure 5), due to fast polymerization on a small amount of stable polyindenyl cations, while in the case of the dimer it increases (Figure 13) because the dimer cation rapidly formed is more stable at low temperature (less secondary reactions).

Assuming complete protonation of the dimer for high values of the R ratio, the corresponding molar absorbance would be 15500 L.mol.$^{-1}$cm^{-1}. A comparison with substituted indanyl cations reported in the literature shows that these cations have molar absorbances in the range of 30000. However, all the cations investigated are substituted in position 1 and not in position 2 as is the dimeric cation. This might explain the difference in ε. Furthermore, the value reported here for high [TfOH] / [dimer] ratios (ε = 15500) is a minimum value since the concentrations of the secondary species absorbing at 405 and 519 nm are not known. They are minimal at low temperatures and at high [TfOH] / [dimer] ratios, but they nevertheless consume a fraction of the dimer.

The absorption band at 519 nm, which appears slowly at high temperature, is a derived cation.

The red colour, which is generally observed during the polymerization of indene when it is carried out under ordinary conditions, has been ascribed to this wavelength[14,15]. Under high purity conditions (e. g. in polymerizations initiated with cumyl chloride and SnCl4 or TiCl4 under vacuum), this coloration is sometimes observed, but it develops slowly and generally after completion of the polymerization, the medium remaining pale yellow during the process, which is generally rapid (seconds or minutes depending on the conditions). Thus the species absorbing at 519 nm should correspond to the cation of a compound resulting from secondary reactions. Prosser and coll. [15] have shown that the allylic anion derived from the dimer of indene through proton abstraction by butyl lithium absorbs in the 470-520 nm range (depending on the nature of the solvent). These authors have assumed that the corresponding cation absorbs at a similar wavelength. In the case of the polymerization of indene, they have proposed for this species a cation formed by hydride abstraction by the growing chain from the penultimate indene unit of a polymer chain having an unsaturated end group (e. g. resulting from transfer). This would explain that in ordinary (non controlled) polymerizations in which transfer reactions take place, this coloration invariably appears, while it is not always visible in controlled polymerizations in which transfer is suppressed or at least reduced. In the present case, the hydride abstraction would involve a molecule of dimer, and would take place concurrently with ionization (scheme 6).

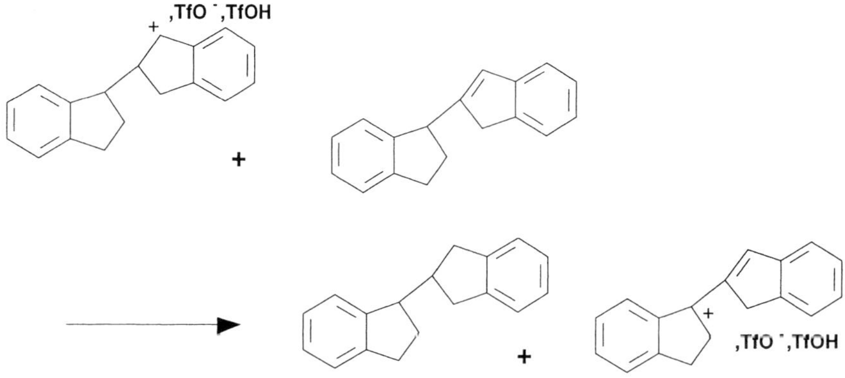

Scheme 6

This would explain the evolution of this peak with the TfOH/dimer ratio at -66°C. As mentioned before, for low values of this ratio (R = 3 and 6) initiation is slow (150 and 50 s) and the unreacted dimer present in the solution can undergo hydride abstraction to yield the species absorbing at 519 nm. For higher values of this ratio, initiation is fast, completed in two seconds, and there is no unreacted dimer available for hydride abstraction, and consequently no important absorption at 519 nm. A small amount of this compound is however formed at the very beginning of the reaction and its concentration remains constant afterwards.

It should also be noted that for high [TfOH] / [dimer] ratios, this maximum remains stable. If the case of a protonation-deprotonation equilibrium leaving a sizeable amount of unsaturated dimer, the hydride transfer would take place, shifting the equilibrium towards deprotonation to yield the species absorbing at 519 nm and this maximum would increase continuously. This apparently happens at high temperatures or at low [TfOH] / [dimer] ratios, in which cases protonation is incomplete.

Concerning the maximum at 405 nm, some authors have assumed that indanyl cations have two maxima. Pittman [28] reported two maxima at 315 nm (ε = 15700 M^{-1} cm^{-1}) and at 400 nm (ε = 6600 M^{-1} cm^{-1}) for the 1,3,3-trimethylindanyl cation and at 301 nm (ε = 14100 M^{-1} cm^{-1}) and 401 nm (ε = 2200 M^{-1} cm^{-1}) for the 1-methylindanyl cation. If the maximum at 405 nm belongs to the indanyl cation, the ratio of the absorbances at 325 and 405 nm should be independent of the reaction conditions. This is approximately verified in the case of the protonation of the dimer at variable temperature, a satisfactory correlation between these two maxima are observed during the first ten seconds of reaction (Figure 14). This would suggest that the maximum at 405 nm might also correspond to the indanyl cation.

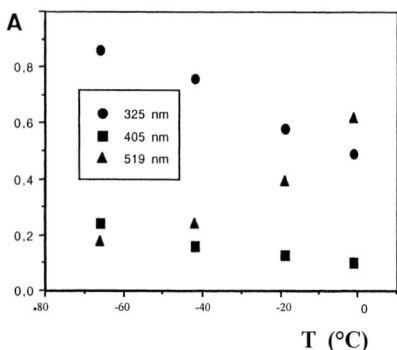

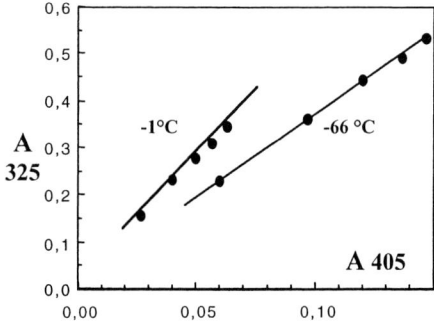

Figure 13 : Protonation of 2-indanyl indene by triflic acid at variable temperature. Evolution of the maximum absorbances with the reaction temperature. [Dimer] = 2.6 10^{-4} mol.L^{-1} ; [TfOH] =4 10^{-3} mol.L^{-1}

Figure 14 : Protonation of 2-indanyl indene by triflic acid at variable temperature. Evolution of the maxima at 405 nm and 325 nm during the first ten seconds of reaction. [Dimer] = 2.6 10^{-4} mol.L^{-1} ; [TfOH] = 4 10^{-3} mol.L^{-1}

Another point to be noticed is the stability of the maximum at 325 nm. This is to be compared to the protonation of styrene in similar conditions [29], in which case a transient absorption at 343 nm and a sideband at 332 nm have been observed, with shorter lifetimes

(250 ms at 0°C; 100 s at -65°C), together with a secondary band at 415 nm which has been assigned to a terminal phenylindanyl cation resulting from indanic cyclization. This reaction does not take place in the case of indene, which may explain the stability of this cation.

CONCLUSION

The experimental observations may be rationalized in the following way:

The absorption in the 318-325 nm range should belong to the indanyl cation and to the corresponding dimer cation. The cation of the growing polyindene chain would absorb at 330 nm. A secondary absorbance at 405 nm may also belong to these cations, but this is not yet clearly demonstrated. The maximum at 519 nm, which appears when unreacted dimer is present in solution, or in the case of the polymerization of indene in conditions favouring chain transfer, results from a side reaction yielding a 1-indenylindane cation.

These indanyl cations are stable at low temperature and still relatively so at -20°C and 0°C. Thus indene, which does not undergo indanic cyclization, yields stable cations and confirms the advantages of this monomer for controlled polymerization.

References

1) L. Thomas, M. Tardi, A. Polton, P. Sigwalt, Macromolecules, 1993, **26**, 4075

2) L. Thomas, A. Polton, M. Tardi, P. Sigwalt, Macromolecules, 1995, **28**, 2105

3) L. Thomas, A. Polton, M. Tardi, P. Sigwalt, Macromolecules, 1992, **25**, 5886

4) P. Sigwalt, K. Matyjaszewski, M. Moreau, Makromol. Chem.,
   Macromol. Symp., **13/14**, 61 (1988)

5) Carbonium ions, G. A. Olah & P. von Schleyer, Interscience vol 1. Interscience, 1968

6) N. C. Deno, C. U. Pittman, J. O. Turner, J. Am. Chem. Soc. , **87**, 2153 (1965)

7) N.C. Deno, P.T. Groves, J. J. Jaruselski, M. N. Lugash,
   J. Am. Chem. Soc., **82**, 4719 (1960)

8) P. Sigwalt, Makromol. Chem., **175**, 1017 (1974)

9) H. Cheradame, A. H. Nguyen, P. Sigwalt, J. Polymer Sci.,
   ACS Symposium, **56**, 335 (1976)

10) A. Leborgne, D. Souverain, G. Sauvet, P. Sigwalt, European Polymer J., **16**, 855 (1980)

11) D. Souverain, A. Leborgne, G. Sauvet, P. Sigwalt, European Polymer J., **16**, 861 (1980)

12) Nguyen Anh Hung, F. Subira, H. Cheradame, P. Sigwalt; Tetrahedron, **34**, 335 (1978)

13) See ref. 5 p. 168

14) A. D. Eckard, A. Ledwith, D. C. Sherrington, Polymer, **12**, 444 (1971)

15) H. J. Prosser and R. N. Young European Polym. J., **8**, 879 (1972)

16) Teng Xu, J. F. Haw, J. Am. Chem. Soc., **116**, 10188 (1994)

17) K. Pitchumani, V. Ramamurphy, Chem. Commun., 2763 (1996)

18) M. De Sorgo, D. C. Pepper, M. Szwarc, Chem. Commun., 419 (1973)

19) T. Kunitake, K. Takarabe, Macromolecules, **12**, 1061 (1979)

20) M. Sawamoto, T. Higashimura, A. Enokida, T. Okubo, Polym. Bull, **2**, 309 (1980)

21) M. Sawamoto, T. Higashimura, Macromolecules, **12**, 581 (1979)

22) T. Kunitake, K. Takarabe, Polymer J., **12**, 245 (1980)

23) T. Kunitake, K. Takarabe, J. Polymer Sci., ACS Symposium, **56**, 33 (1976)

24) M. Villesange, A. Rives, C. Bunel, J. -P. Vairon, M. Froyen, M. Van Beylen,
    A. Persoons, Makromol. Chem., Macromol. Symp., **47**, 271 (1991).

25) J. -P. Vairon, A. Rives, C. Bunel, Makromol Chem., Macromol Symp., **60**, 97 (1992)

26) B. Charleux, A. Rives, J. -P. Vairon, K. Matyjaszewski Macromolecules, **29**, 5777 (1996)

27) B. Charleux, A. Rives, J. -P. Vairon, K. Matyjaszewski, Macromolecules, **31**, 2403 (1998)

28) J. U. Pittman, Ph.D. thesis, Pennsylvania State University, 1964. University Microfilms,
    Inc., Ann Arbor, Mich., N° 65-6762, p. 91-98, p. 169

29) J. -P. Vairon, B. Charleux, M. Moreau, Ionic polymerization and related processes
    Ed. J. E. Puskas, Kluver Acad Pub., 177 (1999)

*Macromol. Symp. **157**, 93–99 (2000)*   93

New Polymer Structures Based on Vinyl Ether Polymerization

Eric J. Goethals,[1] Wouter Reyntjens,[1] Xiaochun Zhang,[1] Beatrice Verdonck,[1] and Ton Loontjens,[2]

[1]University of Ghent, Department of Organic Chemistry, Polymer Division, Krijgslaan 281, 9000 Ghent, Belgium

[2]DSM Central Research Laboratories, P.O. Box 18, 6160MD Geleen, Netherlands

SUMMARY: Sequential living cationic polymerization of octadecyl vinyl ether (ODVE) and methyl vinyl ether (MVE) was used for the preparation of amphiphilic ABA-type block copolymers. The polymerization of ODVE was initiated with the trimethyl silyl iodide/1,1,3,3-tetramethoxy propane/ZnI_2 system at 0°C in toluene. The living bifunctional polyODVE thus obtained was used as initiator for the polymerization of MVE. Below the LCST of polyMVE (37°C), the copolymers are amphiphiles. Above the LCST of polyMVE, the polyMVE-blocks become hydrophobic and the amphiphilic nature of the block copolymer is lost. This was demonstrated by using the block copolymers as emulsifiers for water/decane mixtures. The emulsions were stable for several hours at room temperature, while the emulsion stability decreased to about 30 seconds at 40°C.

PolyMVE-α,ω-bis-methacrylates were obtained by end-capping of living bifunctional polyMVE with 2-hydroxyethyl methacrylate (HEMA). Copolymerization of these bis-macromers with HEMA leads to segmented networks. The networks showed a reversible swelling/deswelling behavior in water as a function of temperature. This is caused by a change of the hydrophilicity of the polyMVE segments in the networks.

Hexa(chloromethyl)melamine, combined with zinc chloride was found to be an efficient hexafunctional initiator for the living cationic polymerization of vinyl ethers. This simple initiating system opens new ways for the synthesis of endgroup-functionalized star-shaped poly(vinyl ethers).

Introduction

The possibility to polymerize vinyl ethers in a living manner has opened the way for the synthesis of a number of new, well-defined polymer structures with tailored properties[1,2]. In this presentation, two examples will be given. First, the

                CCC 1022-1360/00/$ 17.50+.50/0

synthesis of block-copolymers containing poly(methyl vinyl ether) (polyMVE) will be described. PolyMVE is a polymer that is water-soluble at low temperatures but becomes insoluble above a "lower critical solution temperature" (LCST), which is situated at appr. 37°C[3]. This property has been used to produce materials with temperature-controlled hydrophilicity.

In a second part, the synthesis of endgroup-functionalized, star-shaped polyVE's with a new hexa-functional initiator system based on hexa(chloro-methyl)melamine (HCMM), will be presented.

Results and discussion

Polymerization of MVE

MVE is a gaseous compound at room temperature. Therefore, a small-scale semi-continuous polymerization procedure leading to well-defined polymers has been elaborated. The system consisted of a reactor, which can be cooled to –40°C by means of an external cooling fluid, provided with a magnetic stirrer in which the solvent (toluene) can be distilled directly. Gaseous MVE, delivered by a pressure bottle, is introduced at a controlled rate by means of a gas flow controller. The initiation system was a combination of an acetal and trimethyl silyl iodide (TMSI) with zinc iodide as activator[4]. The sequence of introduction of the reagents was as follows : 1) solvent, 2) MVE to reach a starting monomer concentration of 0.4mol./l, 3) the acetal, 4) TMSI, 5) zinc iodide. Then, monomer addition is continued at a known flow rate until the desired ratio [VE]/[In] is reached. After the monomer addition has been stopped, the polymerization is continued until all monomer has been consumed and is terminated by addition of an alcohol and triethyl amine.

The reaction conditions leading to controlled polymerizations using this reaction system have been studied. It was found that a temperature of -5°C, an initiator to activator ratio of 10/1 and an addition rate of monomer corresponding to 1.16 mol/hr, leads to polymers with predictable molecular weights and narrow molecular weight distributions. Some typical results are shown in Table 1.

Table 1. Living cationic polymerization of MVE at $-5°C$ with semi-continuous monomer addition[a]

$[ZnI_2]$ (mmol/l)	$[I_0]$ (mmol/l)	$[M_{tot}]/[I_0]$	$\overline{M}_n$ (theo)[b]	$\overline{M}_n$ (GPC)[c]	$\overline{M}_w/\overline{M}_n$
2.08	8.41	201	11700	10800	1.15
2.74	10.9	153	8900	7800	1.14

a) solvent: toluene; initiator: TMP, $[I_0]$: initiator concentration, $[M]_{tot}$: total monomer concentration added

b) $\overline{M}_n$ (theo) = $73+58*[M_0]/[I_0]$

c) $\overline{M}_n$ (GPC) calibrated with polystyrene standards

It can be seen that, for molecular weights up to 10,000, the observed values are close to the ones calculated for living polymerization with quantitative initiation. The polydispersities are generally below 1.2.

Synthesis of amphiphilic block copolymers from ODVE and MVE

ODVE was polymerized at $-5°C$ in toluene using the bifunctional initiator system 1,1,3,3-tetramethoxy propane (TMP)/TMSI/ZnI_2 system, as described earlier[5, 6]. After completion of the polymerization, MVE was introduced at a controlled rate to produce an ABA block copolymer with the following characteristics:

$\overline{M}_n$ of the central ODVE block: 3500 ($\overline{M}_w/\overline{M}_n$=1.20)

$\overline{M}_n$ of the ABA copolymer: 7200 ($\overline{M}_w/\overline{M}_n$=1.20)

To study the emulsifying properties of the ABA copolymer an ASTM method was used[7]. In this method, equal volumes of water and an organic solvent are mixed thoroughly with a blender and the thus formed emulsion is transferred to a graduated cylinder. The time necessary for 10% demixing is taken as a measure for the emulsion stability. In the present tests, decane was selected as the organic phase. It was already shown earlier, that polyODVE-MVE block copolymer behaves as an amphiphilic polymer, which has emulsifying properties at room temperature[8]. Above the LCST of polyMVE (i.e. 37°C), the block copolymer is

96

no longer amphiphilic and thus looses its emulsifying properties. This was indeed observed as shown in Fig.1 where the 10% demixing times of water – decane emulsions are plotted for different block copolymer concentrations at two temperatures. This figure demonstrates that the emulsifying properties have completely been lost at 40°C.

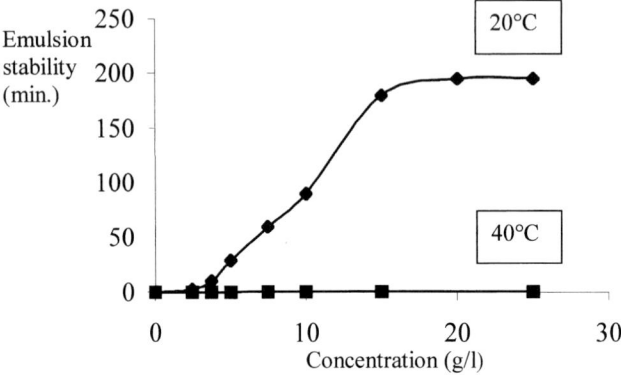

Fig. 1: Temperature dependence of emulsion stability of water-decane (50/50) mixtures as a function of block-copolymer concentration. Emulsion stability measured as the time necessary for 10% demixing.

Synthesis of segmented polymer networks containing polyMVE segments

When bifunctionally living polyMVE is terminated with 2-hydroxyethyl methacrylate (HEMA), the corresponding α,ω-bis-methacrylate is formed:

This bis-macromonomer has been copolymerized with HEMA. The resulting segmented polymer networks are hydrophilic and show considerable swelling in water at 20°C. Increase of the temperature above the LCST of polyMVE results in a drastic decrease in swelling. This is illustrated by Fig.2, in which repeated swelling – shrinking in water, of a copolymer network containing 40% of polyMVE, by repeated cooling – heating, is shown. PolyMVE networks have been prepared earlier by γ-ray induced cross-linking of linear polyMVE[9, 10]. We believe that the use of the bis-macromonomer as starting material for the preparation of networks is more versatile and allows a control of properties by copolymerization with various comonomers.

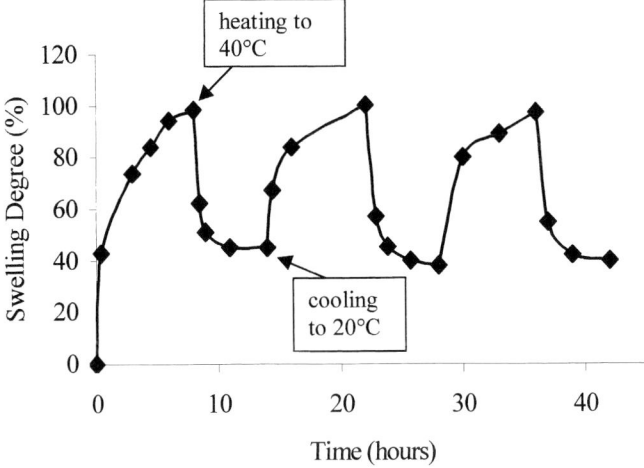

Fig. 2: Swelling-deswelling kinetics of polyMVE-polyHEMA segmented networks in water. Network composition: polyMVE/polyHEMA = 40/60 (mass).

Synthesis of star-shaped polyVE's using a new multifunctional initiator

Hexa(methoxymethyl)melamine (HMMM) reacts with boron trichloride to form hexa(chloromethyl)melamine (HCMM). It was found that HCMM, in combination with zinc chloride, was able to initiate the polymerization of VE's. The polymerizations were quantitative and were terminated with an alcohol in the presence of triethyl amine. The proposed reaction scheme is as follows:

The presence of the aromatic triazine ring in the polymer was evidenced by the observation that, in GPC analysis, a UV-absorption was detected at the elution volume of the polymer. The molecular weight distributions of the obtained polymers were relatively narrow (1.1 – 1.3) and the ratios [polymer protons]/[endgroup protons] in the H-NMR-spectrum, are in agreement with the formation of one endgroup for each chloromethyl unit in the initiator system, suggesting that a hexa-functional living polymerization has taken place. The molecular weights measured by GPC are appr. 30% lower than the values, calculated assuming the formation of a hexa-armed star-shaped polymer, which is in agreement with the fact that star-shaped polymers have a smaller hydrodynamic volume compared with that of their linear analogues of the same molecular weight.

Table 2 gives a survey of some characteristics of the polymerization and the end-products obtained with isobutyl vinyl ether (IBVE).

Table 2. Polymerization of IBVE with HCMM/ZnCl$_2$ at 0°C in toluene

$\overline{M}_n(calc.)^{a)}$	$\overline{M}_n (NMR)^{b)}$	$\overline{M}_n (GPC)^{c)}$	$\overline{M}_w/\overline{M}_n$
7200	6790	4530	1.28
14890	13410	9870	1.32
19900	16490	12910	1.31
31610	29220	22400	1.28

a) $\overline{M}_n$ (calc.)=([M]/[I$_0$])*100 + 390 (MW of HMMM)

b) $\overline{M}_n$ (NMR) from NMR peak ratio of CH$_3$ group of IBVE and acetal end group

c) $\overline{M}_n$ (GPC) based on polystyrene

When the polymerization was terminated by HEMA, a polymer with methacrylate endgroups was obtained. This multi-macromonomer was converted into an insoluble polymer network by exposure to air. This observation confirms that each polymer contains several endgroups.

References

1. M. Sawamoto, *Trends pol. Sci.*, **1**, 111 (1993)
2. "Cationic polymerization", K. Matyjaszewski, Ed., Marcel Dekker inc., New York (1996)
3. Moerkerke R., Mondelaers W., Schacht E., Dušek K., Šolc K., *Macromolecules*, **31**, 2223 (1998)
4. S.S. Lievens, E.J. Goethals, *Polym. Int.*, **41**, 277 (1996)
5. E.J. Goethals, W. G. Reyntjens, S.S. Lievens, *Macromol. Symp.*, **132**, 57 (1998)
6. V. Bennevault, F. Larrue, A. Deffieux, *Macromol. Chem. Phys.*, **196**, 3075 (1995)
7. ASTM-D 244 (1977): SAS-A 3742
8. S.S. Lievens, E.J. Goethals, *Polym. Int.*, **41**, 437 (1996)
9. B.G. Kabra, M.K. Akhtar, S.H. Gehrke, *Polym.*, **33**, 990 (1992)
10. R. Kishi, H. Ichijo, O. Hirasa, *J. Intel. Mat. Syst. Struct.*, **4**, 533 (1993)

Cationic Macromolecular Engineering Via Furan Derivatives

Rudolf Faust

Polymer Science Program, Department of Chemistry, University of Massachusetts
Lowell, One University Avenue, Lowell, MA 01854, USA

SUMMARY: General methodologies for the synthesis of linear and star-block
copolymers and functional polymers using furan derivatives are reviewed, based on
a study of addition reactions of 2-substituted furans to living polyisobutylene
(PIB$^+$). Rapid and quantitative mono-addition of 2-alkyl (and aryl)furans (2-R-Fu),
a new class of non-(homo)polymerizable monomers, to PIB$^+$ has been observed in
conjunction with TiCl$_4$ as Lewis acid in hexane/CH$_2$Cl$_2$ or CH$_3$Cl 60/40 (v/v) at −
80 °C and with BCl$_3$ in CH$_3$Cl at −40 °C, in the presence of proton trap. The
resulting stable allylic cation (P̂IB-Fu$^+$-R) was found to be an efficient initiating
species for the polymerization of methyl vinyl ether (MeVE). Using 2,5-*bis*[1-
furanyl)-1-methylethyl]-furan, coupling of living PIB was found to be rapid and
quantitative in hexanes(Hex)/CH$_3$Cl 60/40 or 40/60 (v/v) solvent mixtures at −80
°C in conjunction with TiCl$_4$, as well as in CH$_3$Cl at −40 °C with BCl$_3$ as Lewis
acid. The kinetics of addition of 2-R-Fu to PIB$^+$ was investigated by UV-vis
spectroscopy. Based on these results and those obtained with substituted 1,1-
diarylethylenes earlier, the absolute rate constants of ionization (k_i), reversible
termination (k_{-i}) and propagation (k_p) in the polymerization of IB have been
determined. By a similar approach the corresponding rate constants have also been
determined for the polymerization of styrene.

Introduction

The importance of non-(homo)polymerizable monomers, such as 1,1-diphenylethylene (DPE)

or 1,1-ditolylethylene (DTE), in cationic macromolecular design and synthesis, arises from

their application in the capping reaction of living cationic polymers. A quantitative mono-

addition of DPE or DTE, to a living chain end, i.e., a capping reaction, results in a stable and

completely ionized cationic living chain end, which has been shown to be well suited for

quantitative end-functionalization with a variety of nucleophiles. The resulting diarylcarbenium

ion was also found to be an efficient initiating species for the polymerization of reactive

monomers. Recently, *bis*-DPE compounds have been successfully used as 'living' coupling

agents for the synthesis of A$_2$B$_2$ star block copolymers. Mechanistic and kinetic aspects of the

capping reactions as well as applications for the synthesis of novel chain-end functionalized

polymers and block copolymers have been reviewed recently.[1] A limitation of using diarylethylenes as capping agents is that quantitative capping can be obtained only under selected conditions. For instance, it is difficult to achieve complete capping of PIB$^+$ with DPE at -40 °C with TiCl$_4$, or even at -80 °C with BCl$_3$, due to the low equilibrium constants of capping/decapping under these conditions. Recently we have discovered a new class of non-(homo)polymerizable monomers; 2-substituted furans as capping agents.

In this paper we report on new methodologies using furan derivatives as capping agents for the synthesis of functional polymers and block copolymers and on kinetic and mechanistic studies.

Results and Discussion

Capping with 2-alkylfurans. Although it has been known that 2-alkylfurans are highly reactive toward electrophilic attack at C–5 position, no attempt has been made on the synthetic utilization of 2-alkylfurans in the living cationic polymerization of vinyl monomers. Our mechanistic studies revealed that 2-alkylfurans, such as 2-methylfuran (2-MeFu), 2-*tert*-butylfuran (2-tBuFu), and 2-phenylfuran (2-PhFu), quantitatively add to living PIB, obtained in Hex/CH$_2$Cl$_2$ or CH$_3$Cl 60/40 (v/v) at -80 °C as shown in Scheme 1.[2] Addition occurs exclusively at C–5 position and a stable tertiary allylic cation is generated at C–2 position which is further stabilized by the neighboring O-atom. The formation of the stable allylic cation (PIB-Fu$^+$-R) was further confirmed by trapping the resulting cation with tributyltin hydride (Bu$_3$SnH). Interestingly, quenching with methanol at −80 °C yielded 2-alkyl-5-PIB-furan, most probably due to the intermediate formation of an acetal, which eliminates methanol.

The stability of PIB-Fu$^+$-R with increasing temperature was studied in Hex/CH$_2$Cl$_2$ 60/40 (v/v) by UV-vis spectroscopy (monitoring the maximum absorption at λ_{max} = 340 nm with R=Ph)[3] or by trapping the remaining cations with Bu$_3$SnH (with R=tBu), followed by the analysis of the product by ^{1}H NMR spectroscopy.[2] Both experiments verified that PIB-Fu$^+$-R is stable up to −40 °C, however it slowly eliminates H$^+$ at -20 °C. Importantly, decapping was not observed at any temperature, indicating that with 2-alkyl(or aryl)furans retro-addition (decapping) is absent or negligible.

Quantitative capping of living PIB with 2-R-Fu could also be achieved using the BCl$_3$/CH$_3$Cl/−40 °C system. Since the capping reaction of living PIB with DPE or DTE was found to be slow and incomplete under these conditions, it appears that 2-alkylfurans are

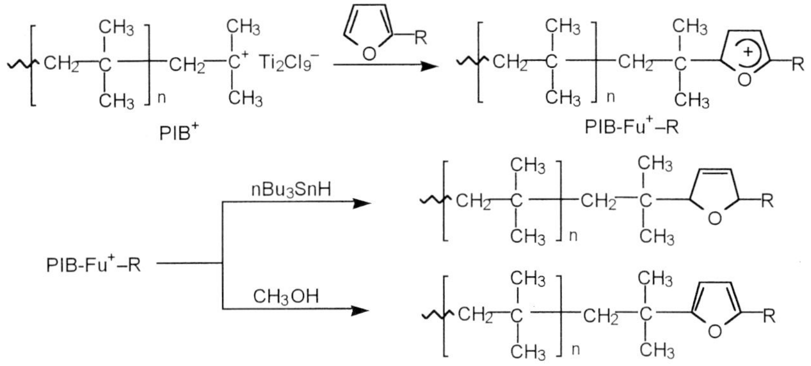

Scheme 1. Capping reaction of living PIB with 2-alkyl(or aryl)furans

more suitable capping agents when the subsequent functionalization or block copolymerization should be carried out at elevated temperature.

Block Copolymerization via Capping with 2-alkylfurans. Upon capping the living PIB chain ends with 2-alkylfurans a stable carbocation is generated which may be used to initiate the polymerization of highly reactive monomers such as vinyl ethers. This concept was tested using 2-tBuFu and 2-MeFu as capping agent for the synthesis of P(IB-b-MeVE) block copolymers by sequential monomer addition. [2] After capping at -80 °C, the Lewis acidity was moderated by the introduction of Ti(OiPr)$_4$ or Ti(OEt)$_4$ followed by the addition of MeVE. The temperature was then raised to 0 °C to polymerize MeVE. Characterization of the products by ^{1}H NMR spectroscopy and GPC indicated that the cross over was not quantitative. By column chromatography the cross over efficiency was determined to be 66% for 2-tBuFu and 75% for 2-MeFu. ^{13}C NMR and DEPT 135 spectra have shown that contrary to expectations, initiation is not from C-2 but from C-4 position. Therefore it appears that PIB-Fu$^+$-R is sterically hindered somewhat which adversely affects the efficiency of initiation of a second monomer.

The Synthesis of Furan Functional PIBs. Recently we reported[4] the quantitative synthesis of furan functionalized PIBs by the reaction of polyisobutenyl diphenylcarbenium ion (PIB-DPE$^+$) with 2-Bu$_3$SnFu. Continuing our investigations we have found that the reaction between 2-Bu$_3$SnFu and living PIB$^+$ (i.e., without DPE capping) in Hex/CH$_3$Cl 60/40 (v/v) in the presence of TiCl$_4$ at -80 °C is also rapid and quantitative. [2] In contrast the functionalization of living PIB with 2-Bu$_3$SnFu in the presence of BCl$_3$ in CH$_3$Cl at -40 °C failed as the PIBCl chain ends remained unreacted even after 15 hours. This may be attributed to transmetallation that gives rise to the formation of BCl$_2$Fu (and nBu$_3$SnCl) which is likely too weak to ionize PIBCl.

Reactions of Furan Functional PIBs. Since furan functional PIB (2-PIB-Fu) can be considered as a polymeric capping agent, 2-PIB-Fu was expected to react with living cationic polymers in coupling reactions. The proof of this concept was obtained in a reaction of 2-PIB-Fu with living PIB$^+$ under conditions similar to capping PIB$^+$ with 2-tBuFu.[2] According to GPC and ^{1}H NMR spectroscopy, close to quantitative coupling of the two PIB chains was achieved in ~3h. The M$_n$ of the product was almost equal to the sum of the M$_n$s of 2-PIB-Fu and living PIB, and the ^{1}H NMR spectrum of the coupled product exhibited a singlet at 5.8ppm since the 2 protons H^3 and H^4 on the ring are chemical shift equivalent.

While the concept of coupling a living cationic polymer with an ω-furan functional polymer has been demonstrated above using living PIB$^+$ and 2-PIB-Fu, more importantly, coupling may also be utilized to obtain *AB* type block copolymers. It is also apparent that ω-furan functionalized polymers may be used as precursor polymers for the synthesis of *ABC* and *AA'B* type three-arm star-block copolymers, where for instance *A* and *A'* represents PIB segments with different molecular weights and *B* and *C* represent chemically different block segments, such as PMeVE or polystyrene. To illustrate this concept, the strategy for the synthesis of *AA'B* type star-block copolymers, where *A*=PIB(1), *A'*=PIB(2) and *B*=PMeVE, is shown in Scheme 2. It involves the coupling reaction of PIB(1) with ω-furan functionality

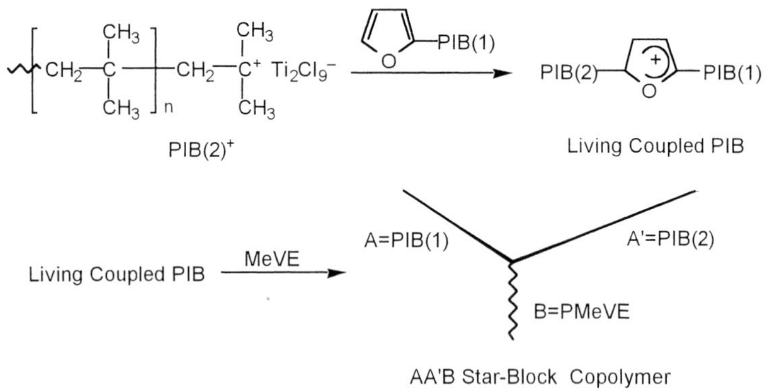

Scheme 2. Synthesis of AA'B tri-arm type star-block copolymers

(*A*) with living PIB(2) of a different molecular weight (*A'*), followed by the chain-ramification polymerization of MeVE to yield the PMeVE block segment (*B*). This scheme was experimentally verified by the synthesis and characterization of PIB-*s*-PIB'-*s*-PMeVE mixed tri-arm star block copolymer.[5] Pure star-block copolymer was obtained upon purification of the crude product, contaminated by a small amount of homo-PIB (incomplete crossover) by column chromatography.

Coupling Reaction of Living Cationic Polymers.   Since 2-alkylfurans add rapidly and quantitatively to living PIB to yield stable tertiary allylic cations, the living coupling reaction of living PIB was also studied using *bis*-furanyl compounds (Chart 1).[6]   Using *bis*-furanyl compounds with a methylene spacer group, such as DMF and FMF, as coupling agents, a higher degree of coupling was obtained with FMF (~85% by ^{1}H NMR) than with DMF (~35% by ^{1}H NMR).   While the products obtained in the capping reaction of living PIB with 2-alkylfurans were colorless, the product obtained in the coupling reaction of living PIB by DMF or FMF exhibited a strong orange color.   This observation indicates the presence of a well-known side reaction, i.e., hydride abstraction at α-position to the ring.

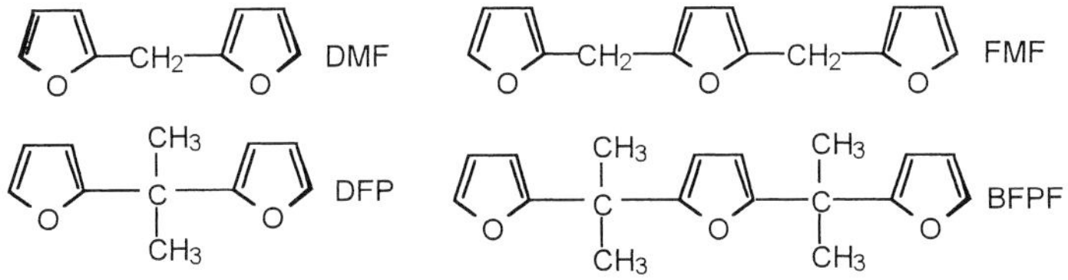

Chart 1. Structures of *bis*-furanyl compounds

This side reaction was circumvented using DFP or BFPF, which lacks hydrogen atoms at α-position to the ring, as a coupling agent.   Using DFP as a coupling agent, however, the coupling reaction was less than quantitative (< 50%) indicating that the reactivity of the second furan ring decreases significantly upon monoaddition.   When BFPF was used as a coupling agent, the coupling was complete within 30 min in Hex/CH$_3$Cl (60/40 or 40/60, v/v) solvent mixtures on the basis of spectroscopic as well as chromatographic analyses.   The ^{1}H NMR spectrum of the final product lacked resonance signals for aromatic group as well as for PIB with a terminal-chloro group, indicating the absence of both monoadduct and unreacted PIB.   Furthermore, the final product exhibited doubled M_n as well as lowered M_w/M_n confirming quantitative coupling of living PIB by BFPF.   Efficient, albeit slow coupling was also achieved with bFPF in CH$_3$Cl at -40 °C in the presence of BCl$_3$.   Unique applications of coupling living PIB with BFPF for the synthesis of hydroxyl, vinyl and dichlorosilyl telechelic PIB with controlled molecular weight have been described.[7]   The *bis*-dimethoxysilyl telechelic PIBs obtained upon quenching with methanol were quantitatively crosslinked by moisture at room temperature in the presence of catalytic amounts of tin(II) 2-ethyl-hexanoate.

Kinetic Aspects of the Capping Reactions. The capping reaction of living cationic polymers with non-(homo)polymerizable monomers comprises of two consecutive reactions as shown in Scheme 3 for PIB and DPE; (i) the ionization of dormant PIB (PIBCl) and (ii) the addition reaction of DPE to ionized PIB (PIB$^+$). Since capping gives rise to a stable and fully ionized cation with high molar absorption coefficient in the UV-vis region, the reaction rate

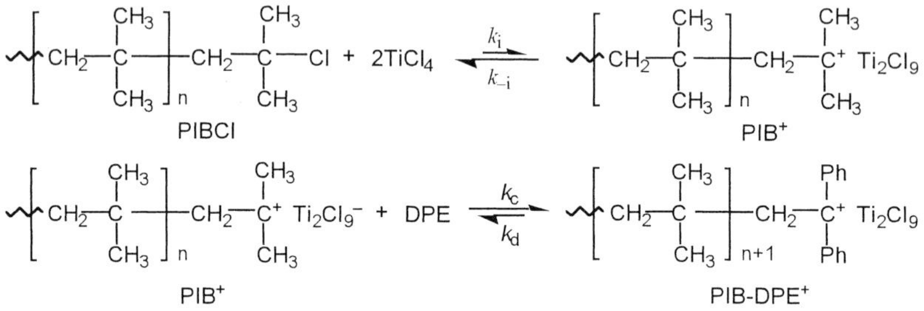

Scheme 3. Ionization equilibrium of living PIB ($K_i = k_i/k_{-i}$) and capping/decapping equilibrium ($K_{cd} = k_c/k_d$)

can be conveniently measured by on-line UV-vis spectroscopy and related to the reactivity of the electrophile and π-nucleophile. We recently studied the addition of 1,1-diarylethylenes to hydrochlorinated IB n-mers, H-[IB]$_n$-Cl (n = 2, 3, 4, 36), in the presence of TiCl$_4$ in hexanes/CH$_3$Cl 60/40 (v/v) at –80 °C using on-line UV-visible spectroscopy.[8] Assuming steady state for PIB$^+$ Ti$_2$Cl$_9^-$, i.e., [PIB$^+$ Ti$_2$Cl$_9^-$]=const., the initial rate of capping for DPE is described by Equation 1.

$$\frac{d}{dt}[PIB-DPE^+Ti_2Cl_9^-] = \frac{k_c k_i[PIBCl][TiCl_4]^2[DPE]}{k_{-i} + k_c[DPE]}$$

Equation 1

From the initial rate of formation of PIB-DPE$^+$ or PIB-DTE$^+$ the apparent rate constants of capping, $k_c K_i$ could be calculated according to Eq. 1, when $k_{-i} \gg k_c$[diarylethylene]. $k_c K_i$ increased with increasing n for the capping with both DPE and DTE: For n = 3, 4, and 36, it was approximately three, four and five times higher, respectively, than for n = 2. Capping with DTE was approximately fifteen times faster than with DPE indicating a much higher reactivity of DTE. With increasing concentration of DTE, a change in the rate determining step from addition of the nucleophile to H-[IB]$_n^+$ ($k_{-i} \gg k_c$[DTE]) to ionization of H-[IB]$_n$-Cl ($k_{-i} \ll$

k_c[DTE]) was observed. The rate constant of ionization, k_i = 6 (n=2), 11 (n=3), and 15 (n=36) M^{-2} s^{-1} could then be calculated for the first time. Comparison of these k_i values with the corresponding k_cK_i values indicated that the observed increase of the apparent rate constant of capping with increasing chain length can be mainly ascribed to a similar increase in k_i and K_i, due to an increase of back-strain with increasing n.

Continuing our investigation on the addition reactions of non-(homo)polymerizable monomers to PIB$^+$ we employed 1,1-*bis*-(4-tert.-butylphenyl)ethylene, (DBE) and 2-PhFu, i.e., π-nucleophiles with increased nucleophilicity. As shown in Figure 1, virtually identical plots have been obtained with DTE, DBE and 2-PhFu, indicating diffusion limited addition with all three nucleophiles. While the nucleophilicity parameters (*N*) have not been determined for diarylethylenes, it has recently been determined for 2-PhFu (*N*=3.6),[9] which in conjunction with the electrophilicity parameter of PIB$^+$ (*E*=7.5)[10] also suggests diffusion controlled addition, according to the linear free energy relationship {*k*=s(*N*+*E*), s≅1}.[10] Using the diffusion controlled second order rate constant of k_c~3x10^9 $M^{-1}s^{-1}$ K_i and k_{-i} could be calculated. On the basis of K_i and the apparent propagation rate constant of IB, obtained under identical conditions, the absolute propagation rate constant k_p (=9x10^8 $M^{-1}s^{-1}$) was determined. A similar approach was used to determine k_p (=5x10^9 $M^{-1}s^{-1}$) for styrene based on data in a recent report.[11] These results are in good agreement with results of the diffusion clock method for IB (k_p=6x10^8 $M^{-1}s^{-1}$) reported earlier[12] and with theoretical predictions based on the linear free energy relationship by Mayr.[10] k_p for styrene however is ~10^6 times higher than that determined using the stopped-flow method.[13] When the addition of the nucleophile to the polymer cation is diffusion limited, and the capped cationic ends do not initiate polymerization of the monomer, a simple competition experiment can also be used to determine k_p. Thus, the polymerization in the presence of capping agent (*C*) will stop short of completion when all chain ends are capped. From the limiting conversion or at low conversion more accurately from the limiting DP_n k_p can be calculated according to Equation 2.

$$DP_{nlim} = \frac{k_p[IB]_o}{k_c[PIBCl]} \ln \frac{[C]_o}{[C]_o - [PIBCl]}$$ Equation 2

The polymerization of IB carried out in the presence of DTE or 2-MeFu (= 3x10^{-3} M= 1.5x [PIBCl]) indeed stopped at ~6% conversion. The ^{1}H NMR spectra of the products verified that all chain ends were capped. From DP_{nlim} k_p = 3 x 10^8 (for IB/2-MeFu), and 2 x 10^8 (for IB/DTE) M^{-1} s^{-1} was calculated.

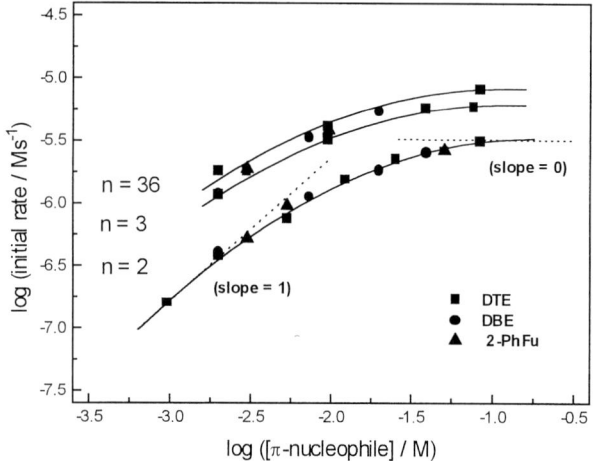

Figure 1. The log initial rate of capping versus log [π-nucleophile] for the capping reaction of hydrochlorinated IB *n*-mers in Hex/MeCl 60/40 (v/v) at –80 °C.

Acknowledgment. The author is grateful to the National Science Foundation for financial support (DMR-9806418 and INT-9512834).

References

1. Y. C. Bae, S. Hadjikyriacou, H. Schlaad, and R. Faust, in *Ionic Polymerization and Related Processes*, Puskas, J. E. Ed. Kluwer Academic Publishers; Dordrecht, Netherlands, (1999)
2. S. Hadjikyriacou and R. Faust, *Macromolecules* **32**, 6393, (1999)
3. Y. Kwon, S. Hadjikyriacou, R. Faust, P. Cabrit, M. Moreau, B. Charleux, and J.-P. Vairon, *Polymer Preprint*, **40**(2), 681 (1999)
4. S. Hadjikyriacou and R. Faust, *Polymer Preprint*, **39**(2) 398 (1998)
5. J. Yun, S. Hadjikyriacou, and R. Faust, *Polymer Preprint*, **40**(2), 1041 (1999)
6. S. Hadjikyriacou and R. Faust, submitted to *Macromolecules*
7. S. Hadjikyriacou, R. Faust and T. Suzuki, submitted to *Macromolecules*
8. H. Schlaad, Y. Kwon, R. Faust and H. Mayr, submitted to *Macromolecules*
9. H. Mayr, private communication
10. H. Mayr, O. Kuhn, M. F. Gotta and M. Patz, *J. Phys. Org. Chem.* **11**, 642 (1998)
11. P. Canale and R. Faust, *Macromolecules* **32**, 2883 (1999)
12. M. Roth and H. Mayr, *Macromolecules* **29**, 6104, (1996)
13. J. P. Vairon, A. Rives, and C. Bunel, *Macromol. Chem., Macromol. Symp.* **60**, 97 (1992)

Macromol. Symp. 157, 109–119 (2000)

FREE RADICAL ACCELERATED CATIONIC POLYMERIZATIONS

James V. Crivello,* Surésh Rajaraman, William A. Mowers, and Saoshi Liu

Department of Chemistry, Rensselaer Polytechnic Institute
Troy, New York 12180-3590 USA

Abstract: The simultaneous photoinitiated cationic polymerizations of epoxides and vinyl ethers in the presence of diaryliodonium salt photoinitiators results in an acceleration of the ring-opening epoxide polymerization and a deceleration of the vinyl ether polymerization. These effects are seen both in mixtures of the two monofunctional monomers as well as in hybrid monomers which bear vinyl ether and epoxide groups in the same molecule. A combination of two mechanisms have been proposed to account for these effects. The reversible conversion of alkoxycarbenium to oxiranium ions results in a two-stage reaction in which first, the epoxide, then the vinyl ether polymerization takes place. Free radical chain induced decomposition of the diaryliodonium salt produces a large incremental number of carbenium ion species which results in the acceleration effect.

INTRODUCTION

Of special interest in this laboratory is the development of epoxide monomers with enhanced reactivity in photoinitiated cationic polymerization. Besides our general interest in the polymerizations of these monomers, they currently have many commercial applications as coatings, inks, adhesives and are potential substrates for such advanced technical areas as fiber optic coatings, optical adhesives, wave guides and in stereolithography. One approach we have recently taken for the design of new, more reactive epoxy monomers has been to incorporate other types of cationically polymerizable functional groups into these monomers. Specifically, we have prepared several such "hybrid" monomers such as allyl and crotyl glycidyl ethers containing both epoxy and vinyl (1-propenyl)[1] ether groups.[2,3] In doing so, we hoped to make use of the higher reactivity of the vinyl ether group in attempt to accelerate the polymerization of the epoxide. A considerable enhancement in the reactivity of the epoxy groups of these two monomer was observed which was ascribed to the presence of the vinyl ether moieties. It was of some interest to determine if the reactivity enhancement displayed by these compounds was unique or whether it was a general phenomenon which could be applied to even more reactive hybrid monomers bearing epoxy and vinyl ether groups. We also wished to examine whether simple mixtures of vinyl ether and epoxide monomers would display similar kinetic characteristics.

CCC 1022-1360/00/$ 17.50+.50/0

RESULTS AND DISCUSSION

Model Compound Studies

The literature contains little information concerning simultaneous epoxide ring-opening and vinyl polymerizations. For this reason, we decided to examine the concurrent photoinitiated cationic polymerizations of several vinyl ether/epoxy monomer pairs. Fourrier transform real-time infrared spectroscopy (RTIR) has proven to be an extremely useful method for monitoring the kinetics of very rapid photopolymerization reactions.[4,5,6,7] Employing this method, we have used 0.5 mol% of **IOC-10** or **IOC-11** as the photoinitiator per reactive functional group in the monomer.

$$n = 10 \text{ (IOC-10)}, \quad n = 11 \text{ (IOC-11)}$$

The progress of both the vinyl ether and epoxide polymerizations were monitored separately and simultaneously by following the decrease in the absorbance of characteristic IR bands at 812 and 1669 cm^{-1} of the respective epoxide and vinyl ether groups.

Figure 1 shows a study, in which the photopolymerizations of the highly reactive cycloaliphatic epoxy monomer **MCEPr** and vinyl (1-propenyl) ether monomer **CP** were conducted.

MCEPr   **CP**

The slopes of the initial portion of the kinetic curves may be used as a measure of the reactivity of these monomers since the $R_p/[M_0]$ values so obtained are directly proportional to the rates of polymerization. While the polymerization of the cycloaliphatic epoxy monomer **MCEPr** is quite rapid ($R_p/[M_0] = 1.0 \times 10^{-2}\text{s}^{-1}$), the polymerization of vinyl ether **CP** is much faster ($R_p/[M_0] = 17 \times 10^{-2}\text{s}^{-1}$).

Next, a 1:1 molar mixture of the above two monomers was prepared and the two monomers simultaneously photopolymerized again using **IOC-10** as the photoinitiator. A markedly different order of reactivity was observed as is shown in Figure 2. In this case, the ring-opening polymerization of the epoxide group displays a marked rate acceleration ($R_p/[M_0] = 3.1 \times 10^{-2}\text{s}^{-1}$). An epoxide acceleration factor, $AF_E = 3.1$ was calculated by taking the ratio of the $R_p/[M_0]$ values for the simultaneous versus the independent polymerizations of **MCEPr**. At the same time, the rate of the vinyl ether polymerization in the mixture was depressed ($R_p/[M_0] = 6.3 \times 10^{-2}\text{s}^{-1}$). A vinyl ether deceleration factor $DF_{VE} = 2.6$ for **CP** was calculated from the ratio of the independent and simultaneous polymerizations. It should also be noted from Figure

2 that the polymerization of the vinyl ether monomer is inhibited until the polymerization of the epoxide is essentially complete.

Synthesis of Hybrid Monomers

Hybrid monomers bearing vinyl ether and epoxy functionalities are readily synthesized using straightforward synthetic methods from available and easily obtained substrates. For example, monomer **CEP** was prepared by the sequence of reactions (eq. 1-3) shown below.

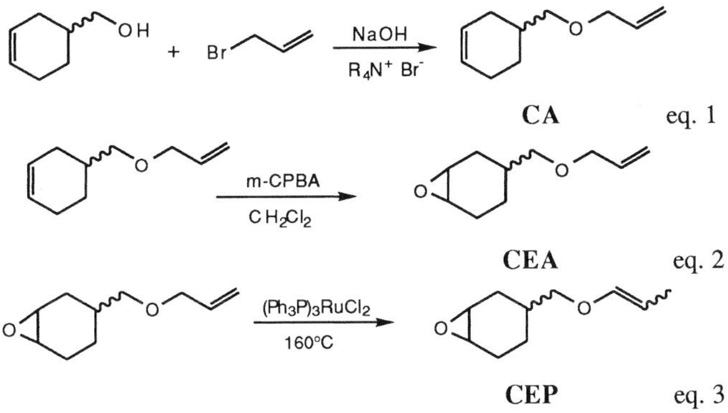

CA eq. 1

CEA eq. 2

CEP eq. 3

CA was prepared in high yield by the Williamson ether synthesis using a phase transfer catalyst (eq. 1). Epoxidation of **CA** using m-chloroperbenzoic acid (eq. 2) takes place regioselectively at the cyclohexene double bond to give **CEA**. Ruthenium catalyzed isomerization of the double bond of **CEA** (eq. 3) proceeds smoothly to give high yields of **CEP**. Similar methodology was employed for the preparation of the three hybrid monomers **MCEP**, **NEP** and **TEP** shown below.

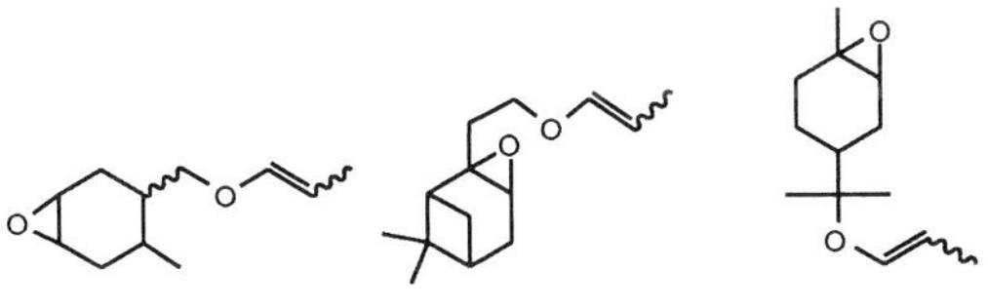

MCEP   **NEP**   **TEP**

The cationic photopolymerization of **CP** is displayed in Figures 3. Analogous behavior is displayed by all the monomers. In each case, the polymerization of the epoxide functional group in the molecule proceeds at a markedly higher rate than the vinyl ether group. This is in direct contrast to the usual situation in which the rate of polymerization of vinyl ethers is typically much faster than that of epoxides. In every case, the polymerization of the epoxide group of the hybrid monomer proceeds to high conversion while that of the vinyl ether group in the same molecule takes place more slowly and to comparatively low conversion. Both polymerizations take place without the usual induction period.

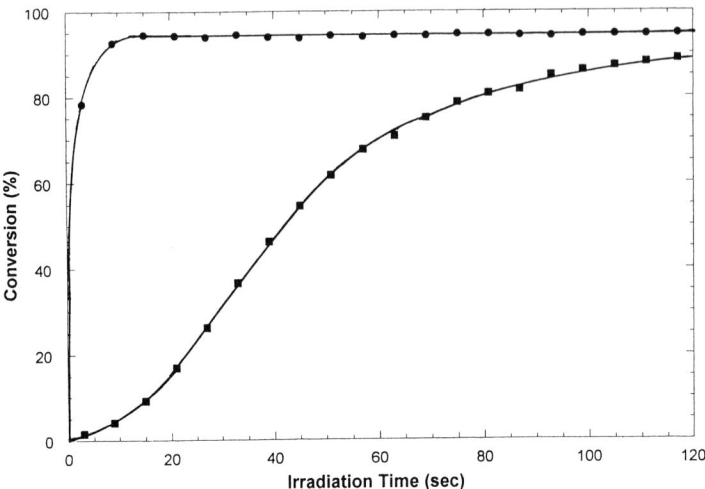

Figure 1. Comparison of the epoxide ring-opening photopolymerization of ■, **MCEPr** and ●, **CP** using 1 mol% **IOC-10** as photoinitiator (light intensity 2150 mJ/cm^2·min).

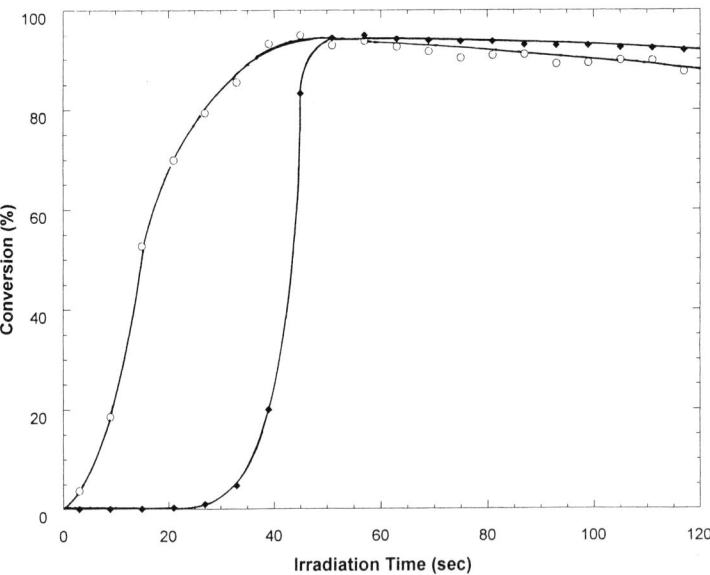

Figure 2. RTIR study of the photopolymerization of a 1:1 molar mixture

of ○, **MCEPr** (epoxide) and ◆, **CP** (1-propenyl ether) using 1 mol% **IOC-10** as photoinitiator (light intensity 2100 mJ/cm^2·min).

Mechanistic Interpretation of the Results

To explain the above phenomena observed with mixed vinyl ether/epoxy monomer systems and with the corresponding hybrid monomers, we propose two mechanistic schemes. The photolysis of diaryliodonium salts as shown in equation 4 takes place by both heterolytic (dominant process) and homolytic cleavages of a carbon-iodine bond and produces a number of reactive species including: radicals, cations and cation-radicals.[8,9,10]

$$Ar_2I^+ \; MtX_n^- \xrightarrow{\; h\nu \;} \left\{ \begin{array}{l} ArI^{\ddot{+}} \; MtX_n^- \; + \; Ar\bullet \\[2mm] ArI^+ \; MtX_n^- \; + \; ArI \end{array} \right\} \longrightarrow HMtX_n$$

eq. 4

Highly reactive aryl cations and aryliodine cation-radicals generated during this photolysis react further with solvents, monomers or impurities to give protonic acids, $HMtX_n$. In this laboratory, we have employed diaryliodonium salts as photoacid generators and used them to photoinitiate the cationic polymerizations of vinyl ethers and epoxides.[11] These two polymerizations proceed respectively, via the formation and propagation of alkoxycarbenium ions and oxiranium ions (eq. 5 and 6).

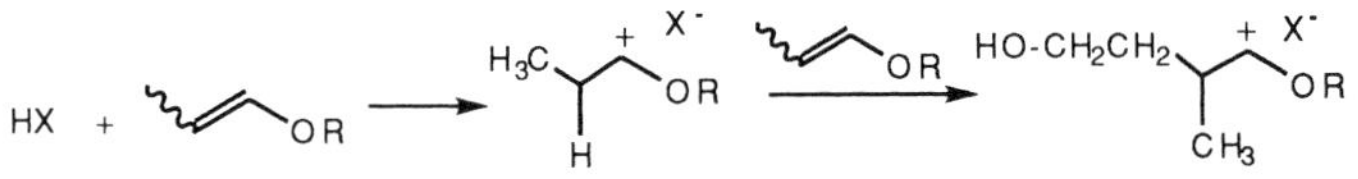

eq. 5

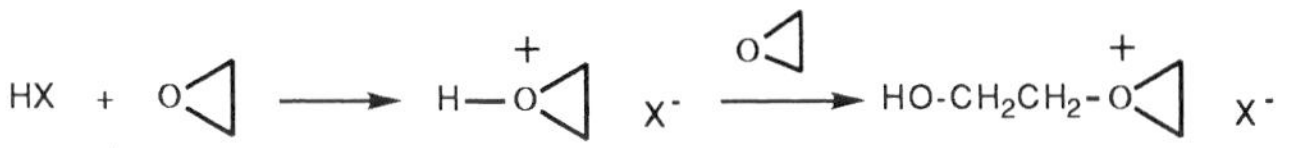

eq. 6

When both reactive species are generated together in the presence of the two different monomers, an equilibrium can be set-up between the two species as shown in Scheme 1.

Scheme 1

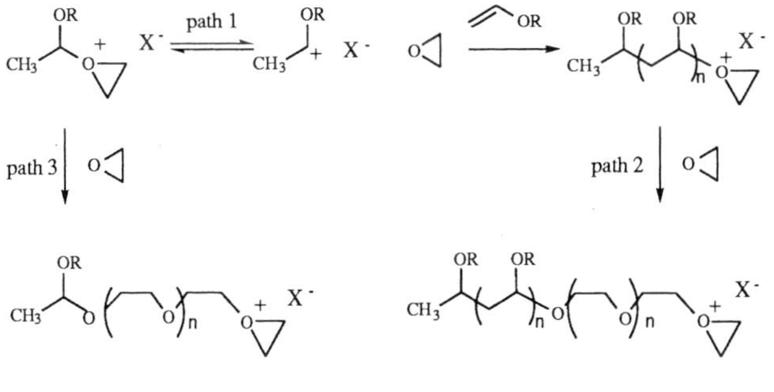

If epoxides of low reactivity such as glycidyl ethers are present, their rate of cationic ring-opening polymerization are comparatively slow. Thus, there will be a small but finite concentration of alkoxycarbenium ions present which can propagate by addition of vinyl ether molecules (path 1). Reversible termination by oxirane molecules and reinitiation may occur many times as vinyl ether polymerization proceeds. Eventually, nucleophilic attack by the oxirane monomer takes place resulting in the start of a polyether chain (path 2) which can no longer equilibrate to form alkoxycarbenium ions and vinyl ether polymerization ceases. In such cases, simultaneous vinyl ether and epoxide polymerizations take place. However, as previously noted, the rate of the vinyl ether polymerization is depressed as compared to the independent polymerization of that monomer. This is due to the rapid conversion of propagating alkoxycarbenium ions to the more stable and less reactive oxiranium ions.

In contrast, if highly reactive epoxide monomers such as those containing epoxycyclohexyl groups are used (path 3), the initially formed oxiranium ions are rapidly converted to growing polyether chains which do not equilibrate with alkoxycarbenium ions. In such cases, all of the polymerization is diverted to epoxide ring-opening and vinyl ether polymerization is delayed until the epoxide concentration falls to a low level at which competition of epoxide monomers for the growing chain end is no longer favorable. In this case, homopolymerization of the two monomers is observed.

The mechanism shown in Scheme 1 bears a striking resemblance to the mechanism proposed by Percec and his coworkers[12,13,14] for the living cationic polymerization of vinyl ethers in the presence dimethylsulfide. These workers have suggested that an alkoxycarbenium ion is in equilibrium with an alkoxymethylsulfonium species and that the propagation and chain transfer reactions are slowed by rapid termination by ion-trapping with dimethylsulfide. For reasons of simplicity, the usual chain transfer and back-biting reactions have not been included in Scheme 1 although the authors acknowledge the presence of these side reactions in the present studies as well. While this mechanism allows us to rationalize the fact that epoxides and vinyl ethers do not undergo any substantial copolymerization, and that in certain cases there is a marked temporal separation between the two homopolymerizations, it does not offer a full explanation for those instances in which an increase in the rate of epoxide ring-opening polymerization in the presence of the vinyl ether monomer is observed. Accordingly, we propose an additional mechanism to explain this latter effect.

As shown in equation 4, free radical species are generated by photolysis of diaryliodonium salts. For this reason, these compounds can be employed to initiate free radical polymerizations. Further, simultaneous, photoinduced free radical and cationic polymerizations can be conducted using diaryliodonium salts.[15,16,17] It is also important to recognize that diaryliodonium salts, due to presence of iodine in the +3 oxidation state, are oxidants.[18] During photolysis, the excited diaryliodonium salt undergoes a formal reduction at the iodine atom. As a result, the solvent, monomer or impurities are oxidized. Diaryliodonium compounds can be reduced in other ways as well. For example, we have employed various photosensitizers (PS) to broaden the spectral response of diaryliodonium salt photoinitiators. The photosensitizers serve as reducing agents in their excited states reducing and the overall mechanism of photosensitization involves an electron transfer or redox process.[19,20]

Hypervalent iodine compounds, including diaryliodonium salts can also be reduced chemically employing a wide variety of chemical reducing agents.[18] In this laboratory, we have employed ascorbate[21] and tin[22] reducing agents together with diaryliodonium salts to conduct redox initiated cationic polymerizations. In two seminal papers, Ledwith, et al.[23,24] demonstrated that electron-rich free radicals produced either by photolysis or by thermolysis are capable of reducing diaryliodonium salts. Yagci and his coworkers[25,26,27] have continued this work in recent years using other types of onium salts. Bi and Neckers[28] have recently reported the use of a dye to photosensitize the visible light cationic polymerization of cyclohexene oxide in the presence of a diaryliodonium salt. These workers reported that the mechanism involved the generation of free radicals by the Norrish Type II abstraction of a hydrogen atom from an amine by the photoexcited dye.

In this laboratory, during the cationic photoinitiated polymerization of 1-propenyl ethers, it was observed that there was a marked rate acceleration and a depression of the induction period when diaryliodonium salts were used as photoinitiators but not when triarylsulfonium salts were employed.[29] In accord with the earlier proposal of Ledwith, et al.,[14,15] we suggested that the observed effects could be explained by the mechanism shown in Scheme 2 involving the abstraction of the α-ether hydrogen atoms by aryl free radicals (eq. 7) followed by the free-radical induced reduction of the diaryliodonium salt by these latter radicals (eq. 8).

Scheme 2

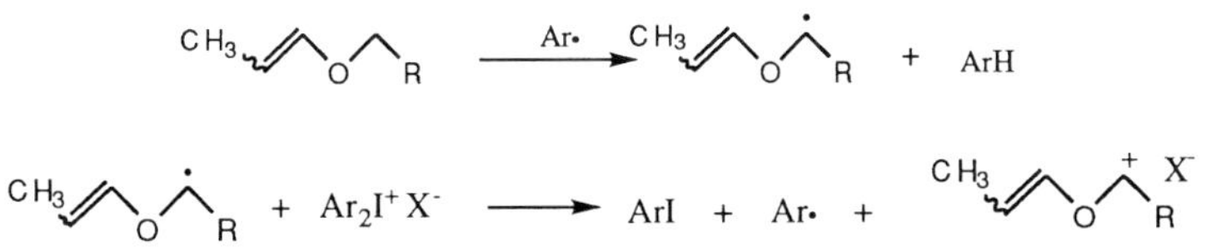

eq. 7

eq. 8

It is also possible, as proposed by Ledwith,[13] that the aryl radicals can add to the vinyl ether double bond to produce an electron-rich and easily oxidized α-ether free radical.

The free radical induced chain reaction represented by equations 7 and 8 results in the dark (i.e. non-photochemical) decomposition of a diaryliodonium salt together with the transformation of an equivalent number of free radicals into cationic centers which can initiate cationic polymerization. Thus, besides the usual direct cationic photoinitiation process, the additional free radical chain induced decomposition of diaryliodonium salts in the presence of vinyl ethers results in the production of substantially more active polymerizing cationic centers and consequently, an increase in rate.

Since the rate acceleration of the epoxide polymerization observed in our studies is a free radical process, it should be inhibited or retarded by free radical traps. We have conducted experiments to verify this hypothesis. Marked reductions in the rates of epoxide polymerizations were observed when nitrobenzene was added to mixed vinyl ether and epoxide systems or to the polymerizations of hybrid monomers. Similar results were noted when these polymerizations were carried out in the presence of oxygen. In addition, the onsets of the vinyl ether polymerization are lengthened in the presence of these retarders. The cationic photopolymerizations of epoxides were not accelerated in the presence of vinyl ethers when triarylsulfonium salt photoinitiators are used. Due to their higher redox potential than diaryliodonium salts, triarylsulfonium salts are not reduced by a-ether radicals.[11] Thus, in this case, cationic polymerization occurs only as a result of the photolysis of the triarylsulfonium salt and without an incremental component due to the free radical induced decomposition of the photoinitiator.

Because the polymerization of epoxide group in hybrid monomers preceeds that of the vinyl ether group in the same molecule, initially a linear polyether polymer is formed. This reaction proceeds to high conversion. When substantially all of the epoxide groups have reacted, the crosslinking polymerization of the pendant vinyl ether groups sets in. This latter polymerization proceeds more slowly since it is essentially a crosslinking reaction. As the reaction takes place, the network becomes more and more highly crosslinked and the reaction rate is retarded by a simultaneous increase in the glass transition temperature. Due to the immobility of the remaining vinyl ether groups, a maximum conversion of only 35% is attained.

CONCLUSIONS

Based on these studies, it can be concluded that the rate of cationic photopolymerization of a given monomer can be influenced not only by such factors as the experimental parameters and the nature of photoinitiator, but also by structural factors within the monomer itself which are remote from site of polymerization. We have taken advantage of this concept to design more reactive monomers by incorporating functional groups which can participate in the free radical induced decomposition of the photoinitiator. In this paper, we have described the synthesis of hybrid monomers which incorporate both vinyl ether and epoxide functional groups in the same molecule. Hybrid monomers **CEP, NEP, MCEP** and **TEP** represent a new class of highly reactive substrates in which the epoxide group polymerizes more rapidly than the vinyl ether group. These new, low-viscosity high-reactivity monomers have many potential applications involving UV curing. In addition, this work also suggests alternative methods by which photoinduced cationic polymerizations of epoxides may be accelerated through rational design of monomers.

EXPERIMENTAL

Photopolymerization Studies Using Real-Time Infrared Spectroscopy (RTIR)

Photopolymerizations of all the monomers were monitored using real-time infrared spectroscopy (RTIR) employing an apparatus and procedures as previously described.[3]

Synthesis of Model Compounds and Monomers

The following experimental procedures are typical of those employed during the course of this investigation for the synthesis of model compounds and hybrid monomers.

Synthesis of (2-Oxapent-4-enyl)cyclohex-3-ene (**CA**)

Into a 500 mL round bottom flask equipped with an overhead stirrer, thermometer and a nitrogen inlet were placed 56.085 g (0.5 mol) of distilled 1,2,3,6-tetrahydrobenzylalcohol, 90.75 g (0.75 mol) of allyl bromide, 100 mL of toluene and 30 g (0.75 mol) of sodium hydroxide. The reaction mixture was stirred at room temperature for 15 minutes. Then, 3 g (0.01 mol) of tetra-n-butylammonium bromide was added and the reaction mixture slowly heated to reflux (65 °C) and maintained at that temperature for eight hours. The reaction mixture was cooled and filtered to remove the sodium bromide which precipitated during the reaction. The filtrate was poured into 500 mL of distilled water, the organic layers were separated and the aqueous layer extracted with fresh toluene. The combined organic layers were washed with three 200 mL portions of distilled water and the organic phase was dried over anhydrous sodium sulfate. Then, the excess allyl bromide and toluene were removed using a rotary evaporator and the reaction mixture subjected to vacuum distillation. The clear liquid distillate amounted to 62.32 g (82 % recovered yield). Fractional vacuum distillation gave pure **CA** with a boiling point of 22 °C at 0.05 mm Hg. Elemental Analysis. Calculated for $C_{10}H_{16}O$: C, 78.90 %; H, 10.59 %. Found: C, 78.82%; H, 10.52%.

Synthesis of (2-Oxapent-3-enyl)cyclohex-3-ene (**CP**)

To 19 g (0.125 mol) of **CA** in a 100 mL round bottom flask equipped with a magnetic stirrer, reflux condenser and a nitrogen inlet were added 0.008 g (0.0075 mmol) of tris(triphenylphosphine)ruthenium(II) dichloride. The reaction mixture was heated at 160 °C for

2 h. The ^{1}H NMR spectrum showed that the bands assigned to the allyl groups (d ppm 5.1-5.4, $CH_2=$; 5.8-6.1, CH= ; 3.8-4.1, CH_2) had been completely replaced by new bands (d ppm 1.58, CH_3; 4.25-4.45, cis-CH_3-C$\underline{H}$= ; 4.65-4.85, trans-CH_3-C$\underline{H}$= ; 5.9-6, cis-CH-O; 6.15-6.3, trans-O-CH) assigned to the 1-propenyl ether groups. Pure **CP** (mixture of *d,l-cis* and *d,l-trans* isomers) was isolated by fractional vacuum distillation (b.p. 156 °C at 20 mm Hg) in 92 % yield. Elemental Analysis. Calculated for $C_{10}H_{16}O$: C, 78.9 %; H, 10.59 %. Found: C, 78.79 %; H, 10.53 %.

Synthesis of (2-Oxapent-4-enyl)-3,4-epoxycyclohexane (**CEA**)

Into a 1 L round bottom flask equipped with an overhead stirrer, an addition funnel and a thermometer were placed 32.3 g of m-chloroperoxybenzoic acid (0.1205 mol) and 300 mL methylene chloride. The flask was cooled to 0-3 °C using an ice bath. **CA** (20 g, 0.1205 mol) in 150 mL dichloromethane was added dropwise so that the temperature did not rise above 10 °C. The addition required approximately 90 minutes. The reaction was allowed to warm to room temperature and then stirred overnight. The reaction mixture was filtered using a Büchner funnel to remove m-chlorobenzoic acid and the filtrate washed with 100 mL quantities of saturated sodium bicarbonate solution until the evolution of carbon dioxide ceased. The organic layer was dried over anhydrous sodium sulfate. Then, the excess methylene chloride was removed using a rotary evaporator and the reaction mixture subjected to vacuum distillation. The volatile clear liquid amounted to 13.96 g (69 % recovered yield). Fractional distillation gave pure **CA** as a mixture of isomers with a boiling point of 30 °C at 0.1 mm Hg. Elemental Analysis. Calculated for $C_{10}H_{16}O_2$: C, 71.39 %; H, 9.59 %. Found: C, 71.36 %; H, 9.55 %.

Synthesis of (2-Oxapent-3-enyl)-3,4-epoxycyclohexane (CEP)

CEA was isomerized using a procedure identical to that used for **CP**. The isomerization reaction was complete after 9 h at 160 °C. A yield of 93 % was obtained. **CEA** (mixture of 8 isomers) had a boiling point of 165 °C at 22 mm Hg. Elemental Analysis. Calculated for $C_{10}H_{16}O_2$: C, 71.39 %; H, 9.59 %. Found: C, 71.39 %; H, 9.53 %.

REFERENCES

1. In this article, we have used the terms "vinyl ether" and "1-propenyl ether" interchangeably.
2. J.V. Crivello and W.-G. Kim, *J. Polym. Sci., Polym. Chem. Ed.*, **32**(9), 1639 (1994).
3. J.V. Crivello and S.S. Liu, *J. Polym. Sci., Polym. Chem. Ed.*, **36**(7) 1176 (1998).
4. C. Decker and K. Moussa, *Makromol. Chem.*, **191**, 963 (1990).
5. C. Decker, *J. Polym. Sci., Polym. Chem. Ed.*, **30**, 913 (1992).
6. U. Müller, *J.Macromol. Sci., Pure and Appl, Chem.*, A**33**, 33, (1996).
7. G. Bradley, R.S. Davidson, G.J. Howgate, C.G.J. Mouillat and P. Turner, *J. Photochem. Photobiol. A: Chem.*, **100**, 109 (1996).
8. R.J. DeVoe, M.R.V. Sahyun, N. Serpone, and D.K. Sharma, *Can. J. Chem.* **65**, 2342 (1987).

9. J.L. Dektar and N.P. Hacker,*J Org Chem* **55**, 639, (1990).
10. J.L. Dektar and N.P. Hacker,*J Org Chem* **56** , 1838 (1991).
11, J.V. Crivello *Ring-Opening Polymerization*, D.J. Brunelle, Editor, 1993, p. 157.
12. J.M. Rodriguez-Parada and V. Percec, *J. Polym Sci., Polym. Chem. Ed.*, **24**, 1363 (1986).
13. V. Percec and D. Tomazos, *Polym. Bull.*, **18**, 239 (1987).
14. V. Percec and M. Lee, *Polym. Preprints*, **32**(1). 212 (1991).
15. J.V. Crivello, *Adv. in Polymer Science, Springer Verlag*, Vol. 62, 1984, p. 1.
16. J.V. Crivello and J.H.W. Lam, *J. Polym. Sci., Polym. Lett. Ed.* **17**, 759 (1979).
17. A.D. Ketley, and J.H. Tsao, *J. Radiat. Curing* **6**(2), 22 (1979).
18. G.A. Olah, *Halonium Ions*, p. 54, New York, Wiley-Interscience 1975.
19. J.V. Crivello, K. Dietliker, *Chemistry & Technology of UV & EB Formulation for Coatings, Inks & Paints, Volume III, Photoinitiators for Free Radical Cationic & Anionic Photopolymerization"*G. Bradley, Editor, 1998, J. Wiley & Sons/SITA Technology Limited, London.
20. J.V. Crivello and J.H.W. Lam, *J. Polym. Sci., Polym. Chem. Ed*, **16**, 2441 (1976).
21. J.V. Crivello and J.H.W. Lam, *J. Polym. Sci., Polym. Chem. Ed.*, **19**, 539 (1981).
22. J.V. Crivello, *Makromol. Chem.*, **184**, 463 (1983).
23. A. Ledwith, *Polymer*, **19**, 1217 (1978).
24. F.A.M. Abdoul-Rasoul, A. Ledwith and Y. Yagci, *Polymer*, **19**, 1219 (1978).
25. Y. Yagci, A. Anen, W. Schnabel, *Macromolecules*, **24**, 4620 (1991).
26. Y. Yagci, *Polym. Commun.* **26**, 7 (1985).
27. A. Onen, Y. Yagci, *J. Makromol, Sci. Chem.* **A27**, 755 (1990).
28. Y. Bi and D.C. Neckers, *Macromolecules*, **27**, 3633 (1994).
29. J.V. Crivello and K.D. Jo, *J. Polym. Sci., Polym. Chem. Ed.* **31**, 2143 (1993).

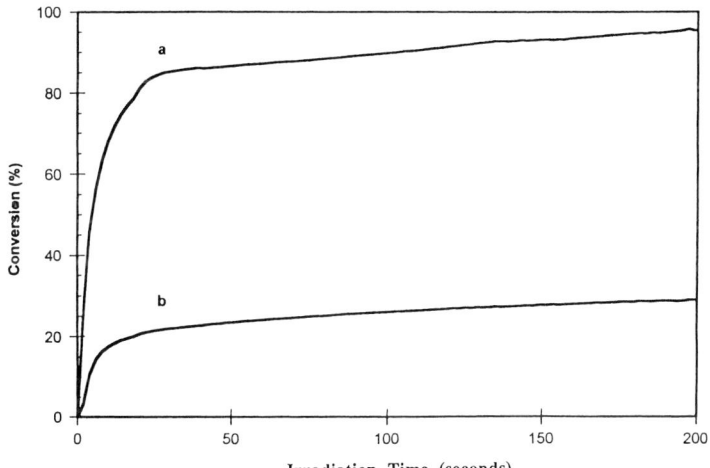

Figure 3. RTIR study of the photoinitiated cationic polymerization of **CEP** using 1 mol% **IOC-11** as photoinitiator. (a), epoxide; (b); 1-propenyl ether (light intensity 2100 mJ/cm^2·min).

SYNTHESIS OF END-FUNCTIONALIZED OLIGOMERS BY ONE-BATCH CATIONIC POLYMERIZATION.

Hung Anh Nguyen, Christine Guis, Hervé Cheradame*
MPI, UMR 7581 CNRS, Université d'Evry, 2 rue H. Dunant, 94320 Thiais, France

Summary

The synthesis of end-functionalized oligomers by cationic polymerization is examined. Special attention is devoted to the problem of achieving this synthesis in a one-batch process, starting from the monomer and an additive which is properly functionalized. The functions of interest are chosen among the pseudohalide functions, azide and isothiocyanate. The functionalizing cationic polymerization of 2-methylpropene was the main system discussed in this context. A brief mention about the behavior of 1,3-pentadiene, isobutylvinylether, and 2-methyl-2-oxazoline is presented. It is shown that depending upon the strength of the Lewis acid and the experimental conditions, the specificity of the functionalization and the possibility of a controlled polymerization can be obtained. The experimental observations can be rationalized by assuming an equilibrium between the active species and the dormant chain ends containing a halide or a pseudohalide. This assumption allows to explain the synthesis of polymers containing two different types of chain termini but corresponding to a narrow molecular weight distribution.

I- Introduction

Our laboratory has been involved in the functionalization of oligomers by cationic polymerization for a long time. More precisely, the goal was a direct functionalization by polymerization from the monomer in a one batch process. Not only, a good control of molecular weight and polydispersity index was looked for, but also the functionalization by a desired chain end such as a pseudohalide was aimed. Among these functions the azide and the isothiocyanate were the more thoroughly investigated, because they were assumed to be sufficiently inert to the conditions of the ordinary cationic polymerization. It will be seen below that this assumption is not always verified. It was expected that, since it was obvious from the work of Kennedy's group that oligomers terminated by a tertiary chloride are obtained in a one-step process starting from a mixture of 2-methylpropene and a benzylic chloride, oligomers could be similarly obtained starting from a mixture of monomer and an initiator suitably functionalized by a pseudohalide function. These functions are interesting because they offer the way to further chemistry such as cycloaddition or transformation into primary amine function for the azide group (1), or addition for the isothiocyanate function (2). It has been published that organic pseudohalides can react with carbenium ions. This reaction in the case of the azide group is known as the Schmidt reaction of the synthesis of the iminium ions (3):

$$R\text{-}C^+,A^- + R'\text{-}C\text{-}N_3 \longrightarrow [R\text{-}C\text{-}NR'=C]^+,A^- + N_2$$

As a consequence of this reaction, the possibility of obtaining the desired functionalization can be questionned. Some work will be described below to clear up this point.

The direct functionalization can be achieved by two ways, either by initiation from a functionalized initiator, or by exchange reaction during polymerization. The first one can be summarized as follows:

$$R\text{-}X + pM \xrightarrow{MX'_n} R\text{-}(\text{-}M\text{-})_p\text{-}X$$

where R-X is a functionalized initiator and M is the monomer. X is a halide or a pseudohalide function. MX'_n is a suitable Lewis acid. The second type of synthesis is represented by the following equation:

$$R\text{-}(\text{-}M\text{-})_{p\text{-}1}\text{-}M^+,C^- + A\text{-}X \longrightarrow R\text{-}(\text{-}M\text{-})_p\text{-}X + A\text{-}C$$

where A-X is a molecule susceptible to exchange its X function with the Lewis acid to give the counteranion of the active species, or with the active species itself. In this study, the A-X molecule was trimethylsilylazide (TMSA) which was purposedly added to a cationically polymerizing system in order to obtain the desired exchange. It can be questionned to know whether the first system above can also be a simple exchange, the R-X molecule being simply a source of -X function. The work described here was also aiming at answering this question, and at determining what would be the easiest way to achieve the desired synthesis.

II- Results and discussion

An important part of the work which is going to be described below was effected with azide group and to a lesser extent with isothiocyanate function. The reason for such choice is that the reactivity of these two pseudohalide functions is suffciently low to be used in similar conditions as the halides. However, to ascertain this point, a preliminary work was carried out to determine at which conditions the cationic polymerization of various monomers could be obtained in the presence of azide function. It was shown that when polymerizing the usual olefins such as 2-methylpropene and styrene the initial concentration of azide function is to be kept lower than that of the Lewis acid (4). The reason for such a condition is to be found in the complexation of the azide function by the Lewis acid which prevents its further reaction with the active species of the cationic polymerization. This is the reason why an excess of free pseudohalide must be avoided, and why with these monomers most of the work was effected in conditions of pseudohalide concentration lower or equal to that of the Lewis acid. In the case of vinyl ethers such as isobutyl vinyl ether, it will be seen below that the condition on the concentration of azide groups is less strict. For instance, the cationic polymerization of isobutylvinylether initiated in the presence of stanic chloride does not seem to be prevented by the presence of an excess of TMSA. It is clear that qualitatively, the reaction of the pseudohalide function with the active species must be a function of the acidity of the active species, and that of the vinylethers is less strong than that of the olefins. Similarly, with heterocyclic monomers, it will be seen below on one example that the cationic polymerization can be carried out in the presence of a large excess of TMSA.

In the various situations encountered in this study, different results were obtained. It is an object of this paper to discuss the results of experiments aiming at the production of functionalized oligomers, in a one-batch process, according to the values measured for the polydispersity index, either a polydispersity index equal or higher than 2, or with a polydispersity index lower than 2. New and old works are presented in order to allow to draw conclusions on the functionalization mechanism.

II-1- Functionalizing cationic polymerization with a polydispersity index equal to two or higher.

A first example which must be described here for the sake of completeness is the polymerization of 2-methylpropene initiated by 1-azido-1methylethylbenzene, also called cumylazide, in the presence of BCl_3 in CH_2Cl_2 at -50°C. Some representative results are shown on Table 1.

Table 1

Polymerization of the system 2-methylpropene (IB)/Cumyl-azide(Cum-N_3)/BCl_3/CH_2Cl_2, at -50°C, [BCl_3] = $1.6.10^{-2}$M, 30 min.

N°	[IB] M	Cum-N_3 M	Yield %	Mn D	f_φ*	f_{N3}**
1	1.1	10^{-2}	30	1600	1.1	0
2	1.1	$1.5.10^{-2}$	80	1700	1.1	0

* Aromatic nuclei content, by NMR spectroscopy and SEC.
**Azide group content, by IR spectroscopy and SEC.

It must be mentionned before the interpretation of these results that initiation takes place from the cumyl azide initiator (φ-C(CH$_3$)$_2$-N$_3$) since without it initiation was not observed in these conditions, and that not only the yield was increasing with the initiator concentration but that all macromolecules contain an aromatic nucleus as shown by the aromatic molar content which is close to one (5). However, the infrared analysis showed that the polymer did not contain any measurable amount of azide group. We conclude from these observations that in our case the following reaction did not take place:

$$PIB^+, BCl_3N_3^- \longrightarrow PIB-N_3 + BCl_3$$

In other words, the complex counteranion $BCl_3N_3^-$ do not release the azide moiety. NMR spectroscopy showed that these polymers contained terminal chloride. Obviously, the interaction between BCl_3 and Cl^- is less strong than with N_3^-. The incorporation of the chloride ion is a termination reaction and explains the yields which were not complete. This first example tells that the functionalization efficiency by the pseudohalide group is governed by the properties of the counteranion in which the pseudohalide anion is incorporated. In this system, the molecular weight distribution does not exhibit any narrowing tendency which demonstrates that termination and transfer occur at random.

A second example is provided by the polymerization of 1,3-pentadiene in the presence of TMSA initiated by $AlCl_3$ in a non polar medium. In this system it was shown that direct initiation is responsible for the active species production (6). Some characteristics of this system are described in Table 2.

Table 2

Polymerization of the system 1,3-pentadiene ([2.3M])/TMSA/AlCl$_3$ ([AlCl$_3$] = 2.3.10^{-2}M)/hexane, at room temperature, reaction time = 2 h.

N°	[TMSA] [AlCl$_3$]	Yield %	Mn soluble fra. D	I* soluble fra.
1	0	93	7000	35
2	1	26	1000	25
3	4	9	600	5

*Polydispersity index.

The first observation is that the presence of TMSA strongly reduced the yield. As expected, a large excess of TMSA was very detrimental (run 3). Analysis of the polymers shows that the polymers produced in the presence of TMSA contained chloride and a small amount of azide groups. To this respect, a comparison of this phenomenology using AlCl$_3$ with an other work carried out on the same monomer in similar conditions but using AlEt$_2$Cl is interesting (7). In this last work, it was shown that with this Lewis acid, initiation did not mainly proceed directly since in the absence of any initiator, the yield was low. It is interesting to notice that in the presence of TMSA no modification was reported in term of yield, molecular weight and polydispersity index. TMSA was inert, while in the presence of an other initiator such as cumyl chloride the polymerization was fast with high conversion and a high proportion of crosslinked product. The same paper (ref. 7) reports that in the presence of bis(azidomethylethyl)benzene (BAMEB), the yield is much better, producing a totally soluble polymer (7). It is clear that in this system, the acidity of the Lewis acid is high enough to induce polymerization from the initiator even in non polar solvent, while it is not able to give any exchange with TMSA. At last, it is important to note that polymerizations carried out in the presence of chlorinated initiators, such as cumyl chloride and AlEt$_2$Cl, produced a high proportion of insoluble material with a slightly better yield than without.

The results of 1,3-pentadiene polymerizations with AlCl$_3$ showed that functionalization constituted a termination reaction, as shown by the reduction of the yield, molecular weight and polydispersity index (Table 2). In agreement with this conclusion, it is worth noting that in the presence of TMSA and AlCl$_3$, the insoluble fraction was strongly reduced (8). A short study was carried out in our laboratory on the point to know whether the polymer could be functionalized by an azide group for the system based on aluminum trichloride. A poly(1,3-pentadiene) was synthesized in the presence of the system BAMEB-AlCl$_3$, and analysed by infrared and ^{1}H NMR spectroscopies (8). It was shown that, after careful cleaning, the polymer contained a small amount of azide group, but no aromatic nucleus. This result is interpreted as showing that azide group functionalization results from an exchange reaction, and not from initiation or transfer from BAMEB. In that case, BAMEB behaves as a simple source of azide groups like TMSA. This conclusion could seem contradictory with the results reported in ref.7 showing initiation from BAMEB. It must be recalled that in the AlCl$_3$-based system initiation proceeds directly, producing by self ionization an ion pair on which the exchange can take place. It is unfortunate that not enough information was available on the characteristics of polymerizations carried out in the presence of AlEtCl$_2$, and this situation prevents to draw more conclusions from the comparison.

In the same context, the polymerization of isobutylvinylether initiated in the presence of TMSA and SnCl$_4$ in polar solvent must be reported (9). Some characteristic results are described in Table 3. In this system, in the absence of TMSA, initiation should be cocatalytic, using residual moisture. The polymerization yield was never complete, showing that some termination reaction occured. Whatsoever, the molecular weight did not appreciably change with TMSA concentration. This observation is difficult to explain taking into account that the oligomers were functionalized by an azide group when the TMSA concentration was high enough. Since the polydispersity index remained around 2, it must be accepted that the mechanism which governs the interruption of chain growth, functionalization, is predominant. Since. the molecular weight did not decrease when TMSA concentration increased, it can be suggested that functionalization is zero order in active species concentration which allows to

Table 3

Polymerization of the system IBVE (0.7 M)/TMSA/SnCl$_4$ (4.7.10^{-2})/CH$_2$Cl$_2$, 20°C, t = 20 min.

N°	[TMSA] M	Yield %	Mn$_{SEC}$* D	I**	f$_{N3}$
1	1.4.10^{-2}	70	1200	2.4	0.3
2	4.1.10^{-2}	70	1300	2.3	0.75
3	6.2.10^{-2}	67	1400	2.5	1.0
4	10.3.10^{-2}	83	1400	2.2	1.1

*In eq. P. St.
**Polydispersity index.

suggest that the functionalization is due to the release of the azide group by the counteranion the chain ends. The primary reaction of TMSA must happen on the Lewis acid itself, and this is why nothing can be assumed on the initiation mechanism. It is also to be noted that a large excess of TMSA does not seem to be detrimental to the active species, since the best polymerization yield is obtained with the highest TMSA concentration. This is probably to be assigned to the fact that in the case of IBVE polymerization the active species are less acidic than in the case of olefins. This example tells that functionalization by a pseudohalide is possible but not with a complete yield, and that this functionalization does not allow a control of the molecular weight. However, this situation meets the requirement of being possible in a one-batch process.

An other example of interest to be discussed in this context is represented by the polymerization of 2-methyl propene in a polar medium (CH$_2$Cl$_2$) at -70°C in the presence of BF$_3$ (5). A summary of the results is shown on Table 4. These results are dealing with two different initiators, the 2,4,4-trimethyl-2-azidopentane (H-TMP-N$_3$) and the 2-phenyl-2-azidopropane (cumylazide). It is interesting to note that the yields are complete when the azide concentration was not higher than the Lewis acid concentration. Run 3 shows that on the contrary, when the azide concentration was higher than that of BF$_3$, a decrease of the yield is observed. About initiation, the cocatalytic mechanism is likely (10). On the other hand, a coinitiation mechanism must be also observed, and this is witnessed by the bimodal distribution in the case of the polymer synthesized in the presence of H-TMP-N$_3$ (run 1). The fact that two different active species are at work in the case of run 1 entails that the functionalization is not the result of a simple transfer occuring on H-TMP-N$_3$.

In the case of initiation on cumylazide, the macromolecules are containing around one aromatic nuclei per chain. The chain ends in the case of run 2 contain at the same time azide groups and unsaturations. It must be concluded that two mechanisms of interruption of chain growth are at work at the same time, the functionalization by the azide group and transfer. These two competing mechanisms help to understand why the polydispersity index is higher than two. The question is to know whether the cocatalytic initiation can also be observed in the case of cumylazide initiation. The high phenyl ring content shows that at least transfer to cumylazide should be very important. Since the azide content is lower than the phenyl ring content, some dehydroazidation or transfer to monomer must occur. However, it must be noted that transfer to monomer is probably negligible because a high phenyl ring content is measured. The fact that the molecular weight is independent of the Lewis acid concentration is an argument in favour of initiation on cumyl azide and of the disappearance of the cocatalytic initiation. The explanation of this interesting characteristic of the use of azide group containing initiators is to be found in the competion between water molecules and cumylazide molecules for complexation by BF$_3$ molecules. It is clear that the reaction leading to carbocations is faster

Table 4

Polymerization of the system 2-methylpropene (IB)/R-N$_3$/BF$_3$/CH$_2$Cl$_2$, at -70°C, t = 50 min.

N°	IB	R-N$_3$	BF$_3$	Yield	Mn	f$_\varphi$*	f$_{N3}$**
	M	M	M	%	D		
1	1.1	H-TMP-N$_3$ $4.8.10^{-2}$ ****	$4.8.10^{-2}$	100	2600***	-	0.45
2	1.1	Cum-N$_3$ $5.0.10^{-2}$ ****	$5.0.10^{-2}$	100	1300	0.8	0.5
3	1.1	Cum-N$_3$ $5.0.10^{-2}$	$1.5.10^{-2}$	80	1300	1	0.8

* Aromatic nuclei content, by NMR spectroscopy and Mn SEC (in eq. pSt).

**Azide group content, by IR spectroscopy and SEC.

***bimodal distribution.

****H-TMP-N$_3$ = tBu-CH$_2$-C(CH$_3$)$_2$-N$_3$, Cumylazide: φ-C(CH$_3$)$_2$-N$_3$

in the case of coinitiation with cumylazide than in the case of the sequence: cocatalyst reacts with the Lewis acid to give the so-called complex acid which in turn reacts with the monomer to give the carbocation. Several systems behave similarly as far as cocatalysis is concerned, and to this respect they are worth of attention due to the increase of selectivity they confer.

The results of run 3 (Table 4) could have been assigned to a predominant transfer to cumyl azide rather than to coinitiation on cumylazide, since the functionalization by aromatic nuclei and by azide group are close to each other. In order to determine the answer to this question, run 3 must be compared to run 2. This last experiment shows that when the Lewis acid concentration is increased, one loses on the side of functionalization. This is due not only to some transfer (unsaturated chain ends and protonic reinitiation on the other end), but also to deshydroazidation since the number of unsaturated chain ends is not equal to the number of aliphatic chain ends. It will be seen farther in this paper that initiation by coinitiation often is the most probable reaction scheme. Whatsoever, it must be concluded that a strong Lewis acid such as BF_3 does not allow a good control of the chain end functionalization, since we have to compromise between yield and functionalization. A comparison with the next system is illustrating.

The polymerization of 2-methylpropene was initiated by BAMEB in CH_2Cl_2 at -50°C. The Lewis acid was titanium tetrachloride. Some typical results are shown on Table 5. Despite some scatter in the results, it is clear that the yield was never complete. In the case of run 2 the lower yield is due to the fact that the azide concentration was higher than that of the Lewis acid. Runs 2 and 3 gave polydispersity indeces around 2. The result of run 1 must be discussed. It is worth noticing that the initiator (bis-1,4-(1-azido-1-methylethyl)benzene, BAMEB) is bifunctional. Consequently, in the case of a predominant step of interruption of chain growth, a polydispersity index of 1.5 is expected. This is exactly what it is observed. Run 4 was carried out in the presence of a small quantity of ditertiobutylpyridine (DtBP), but with the same conditions as run 3. It can be seen that in term of yield, the results are the same as without proton trap (run 3). However, the functionalization is much lower. This is to be assigned to the fact that DtBP interacts with the active species and modifies their reactivity. This effect was noticed several times in our laboratory. It was also found that all polymers contained chloride. It is worth mentioning that in methylene chloride the solubility of the poly(2-methyl propene) is limited. If the molecular weight obtained in methylene chloride

Table 5
Polymerization of the system $IB/BAMEB/TiCl_4/CH_2Cl_2$, at -50°C, t = 50 min

N°	IB M	BAMEB M ***	[TiCl_4] M	Yield %	Mn D	f_{N3}	I*
1	1.1	1.10^{-2}	2.10^{-2}	92	6200	1.1	1.5
2	1.0	1.10^{-2}	1.10^{-2}	56	5500	0.5	1.8
3	1.0	1.10^{-2}	6.10^{-2}	80	5000	1.6	2
4**	1.2	1.10^{-2}	7.10^{-2}	74	6900	0.8	-

*Polydispersity index.
**In the presence of $6.10^{-3}M$ of DtBP.
***BAMEB = $N_3-C(CH_3)_2-\varphi-C(CH_3)_2-N_3$

would only result from solubility characteristics, the comparison between the results quoted on Table 5 and Table 1 shows that this is not the case, since at the same temperature (-50°C) the molecular weight is different.

A brief study was devoted to the study of small variations of the reaction conditions to see whether some improvement could be found. Various initiators were used, i.e. cumyl azide or bis-1,4-(1-azido-1methylethyl)benzene (BAMEB). The results are shown in the next table

6 (runs 1-4 with cumylazide and run 5 with BAMEB). At -60°C, the yield using $TiCl_4$ is not better than with BF_3 at -70°C.

Run 4 demonstrates once more the detrimental effect of DtBP on the functionalization. On the other hand it is interesting to note that the best functionalization was obtained for a one-to-one ratio of the initiator concentration to that of the Lewis acid (run 2). This result allows to choose among the two possible mechanisms of functionalization, predominant transfer on the azide molecule or coinitiation with the organic azide. Since the theoretical molecular weight assuming one aromatic nucleus per chain is always lower than the experimental one and since the used organic azide (based on the azide functionalization) goes through a maximum for the one-to-one ratio, these observations strongly suggest that the main role of cumyl azide is to be ionized by reaction with the Lewis acid, i.e. to behave as an initiator. Functionalization efficiency is slightly better with BAMEB than with Cum-N_3.

Table 6

Polymerization in the system 2-methylpropene (1.1 M)/
Cum-N_3(runs 1-4) or BAMEB(run5) /$TiCl_4$ (0.05 M)/CH_2Cl_2, at -60°C, t = 30 min.

N°	[-N_3]/[$TiCl_4$]	Yield %	Mn_{th}* D	Mn_{SEC}** D	f_{N3}***	Used N_3 %
1	1.7	90	725	3000	0.5	12
2	1	80	1400	2200	0.6	26
3	0.5	100	2500	3500	0.1	7
4****	1	100	1400	2500	0.2	10
5	1	80	1400	3000	1.1	36

*Assuming one cumyl-N_3 molecule per chain (no transfer).
**In eq. p.St.
***determined by IR spectroscopy.
****Experiment carried out in the presence of DtBP (10^{-3} M).

Experiment N°5, Table 6, gave a polymer having a polydispersity index of 1.5. Part of the chain ends contained chloride, since a -Cl functionality of 0.13 was measured by NMR spectroscopy. It also contained terminal double bonds, their functionality being 0.6 also measured by ^1H NMR spectroscopy.

It is interesting to notice that the results of the experiment 5 in term of yield and functionalization quoted on table 6 is practically the same as that of run 2 carried out in the same conditions of azide group concentration but with a monofunctional initiator. It also shows that the interruption of chain growth occurs at random, the polydispersity index of the polymer 5 being equal to 1.5. The comparison of the same experiment with the run 3 of table 5 carried out in similar conditions does not give hopes to get a definite improvement since a slightly lower polydispersity index is observed but with a slightly lower functionalization. Compared to the results of Table 4, the use of a this new set of conditions (Lewis acid and temperature) does not allow a better functionalization, and, besides tranfer, this can be assigned at least partly to the termination by chloride, which cannot happens with BF_3.

In order to expand the scope of this chemistry, some experiments were carried out with another pseudohalide, the isothiocanate group -NCS. It was first determined whether an alkyl isothiocyanate could initiate the polymerization of 2-methylpropene. The next table 7 describes one experiment with again 2-methylpropene and AlEt$_2$Cl which is a rather weak Lewis acid. It was verified that in CH$_2$Cl$_2$ at -50°C it cannot initiate the polymerization of 2-methylpropene while it does when 2,4,4-trimethyl-2-isothiocyanatopentane (H-TMP-NCS) is present. The polymer was analysed by ^{1}H NMR and infrared spectroscopies, and the surprise came from the much higher content of thiocyanate function than that of isothiocyanate. It has been shown that isomerization of alkyl thiocyanates to isothiocyanates could be obtained, while not quantitatively, and that the rates of isomerization increased in the order of increasing stability of the carbocation derived from the alkyl group atached to the sulfur atom (11). Consequently, at equilibrium for the interchange between the two functions, the isothiocyanate must be the major function, while its relative concentration may depend to some extent upon the reaction conditions. The above observation of the polymer functionalization is in tune with the higher nucleophilicity of the sulfur atom and the higher basicity of the nitrogen atom of the thiocyanate ion. The thermodynamic equilibrium slowly establishes while polymerization proceeds.

For this polymer, the polydispersity index was low, but it was due to the low polymerization degree. The high unsaturation content shows that transfer was extensively occuring and is in tune with the preceding conclusion that the system was neither living nor controlled. It is worth recalling here that the same kind of functionalization was observed

Table 7

Polymerization of the system 2-methylpropene (IB = 0.91M)/
H-TMP-NCS/AlEt$_2$Cl/CH$_2$Cl$_2$, at -50°C, t = 50 min.*

N°	[H-TMP-NCS] M	[AlEt$_2$Cl] M	Yield %	Mn**	f$_{SCN}$	f$_{NCS}$	f$_=$***
1	0.04	0.08	54	500	0.22	0.04	0.7

*H-TMP-NCS = tBu-CH$_2$-C(CH$_3$)$_2$-NCS

**By NMR spectroscopy, based on the t-Bu chain-ends content.

***Unsaturation content, number of double bonds per chain.

when the same type of 2-methylpropene polymerization (2-methylpropene/H-TMP-NCS/CH$_2$Cl$_2$) was tried at -50°C in the presence of ditertiobutylmethylpyridine (12). Another surprise was found in the fact that the system 2-methylpropene/H-TMP-NCS/TiCl$_4$/CH$_2$Cl$_2$ at -50°C, where the Lewis acid concentration was twice the concentration of the organic isothiocyanate, did not give functionalization, H-TMP-NCS being recovered untouched (13). It can be surprising at first sight that a strong Lewis acid (TiCl$_4$) apparently does not interact with an alkyisothiocyanate, while a weaker one such as AlEt$_2$Cl does. It can be suggested that the sulfur atom of the H-TMP-NCS is more accessible to the aluminum atom of AlEt$_2$Cl than to the titanium atom of TiCl$_4$. However, a comparison with another run where the concentration of TiCl$_4$ was equal to that of the same isothiocyanate gave a partly functionalized product. It shows that an excess of TiCl$_4$ allows a polymerization which can polymerize without interaction with the isothiocyanate groups, this polymerization being initiated by direct initiation (13). To this respect, the situation seems to be less interesting that the situation which prevails in the case of azide group. On the other hand, the study of the model system Cumyl-NCS/TMP1/TiCl₄/CH₂Cl₂ at -50°C with a Lewis acid concentration twice that of the cumyl-NCS showed the formation of H-TMP-Cl and H-TMP-NCS. This shows that this isothiocyanate of the benzylic type interacts more readily with TiCl₄ than the alkylisothiocyanate.

II-2- Functionalizing cationic polymerization with a polydispersity index lower than two.

If a polydispersity index lower than 2, and of course as low as possible, is to be preferred, transfer and termination must be eliminated, which means that living system is required, or at least a system of « controlled polymerization ». However, if the termination by azide ion is a true termination low polydispersity indeces should not be possible. An investigation of this point was effected with two different monomers and two pseudohalides, the azide and the isothiocyanate groups. A review of some systems investigated in our laboratory provides the answer to this question.

The first one is a system in which the functionalization of poly(2-methylpropene) is carried out by exchange. Table 8 gives some informations.

Table 8
Polymerization of the system 2-methylpropene/TMSA/
TiCl₄/various solvents, at -70°C, t = 2 h.

N°	[IB] M	[TMSA] M	[TiCl₄] M	Solvent % (v/v)	Yield %	Mn_{SEC}** D	I*	f_{N3}	f_{Cl}
1	1.0	0.13	0.13	CH₂Cl₂ 67 % hexane 33 %	93	7900	1.4	0.4	0.3
2	1.0	0.13	0.13	CH₂Cl₂ 50 % hexane 50 %	26	8100	1.3	0.7	-

*Polydispersity index.
**In equivalent pSt.

In this system, two solvent mixtures were tried, CH₂Cl₂ with two different contents of hexane. If the first mixture gave a rather high yield, the second one gave a much lower yield of the same molecular weight but with higher azide group content. Polymers contained chloride and a small amount of terminal unsaturation. The rather low polydispersity indeces allow to raise an interesting question. If the functionalization is obtained by exchange after polymerization, and if there is some termination reaction, the polydispersity index should not be lower than 2 since the 2-methylpropene is not living in these conditions. The simplest explanation for this low polydispersity index is that functionalization is not effected after polymerization but during polymerization and that this functionalization is not a true termination. Since all polymers contain chloride function, it is proposed that terminal chlorinated chain ends are exchanging with azide groups, so that these groups are also a kind of dormant species. It will be seen below that this proposal can account for our most important observations.

Since at -70°C the system seems to give low polydispersity index, the same type of functionalization by exchange was tried in pure CH_2Cl_2 as described in Table 9. Run 1 shows that the yield was complete in these conditions. It is worth noting here that in this system initiation is due to direct initiation and probably to residual cocatalysis (14).

Table 9
Polymerization of the system 2-methylpropene (IB)/TMSA/
$TiCl_4/CH_2Cl_2$, at -70°C, t = 2 h.

N°	[IB] M	[TMSA] M	[TiCl₄] M	Solvent	Yield %	Mn$_{SEC}$** D	I*	f$_{N3}$	f$_{Cl}$
1	1.2	0.14	0.14	CH_2Cl_2	100	6300	1.3	0.6	0.3
2	0.4	0.03	0.03	CH_2Cl_2	100	3000	1.8	0.4	0.3

*Polydispersity index.
**In equivalent pSt.

As for the chain ends of the polymer corresponding to run 1, most of the macromolecules are terminated by either an azide group or a chloride group. The low polydispersity index indicates that all macromolecules could grow and that the terminal functionalization was not a termination reaction occuring at random. Run 2 deserves some comments. This experiment was approximately the same as run 1 but using a dilution factor of around 4. ^{1}H NMR analysis showed that the polymer contained a noticeable amount of terminal unsaturation, which showed the existence of transfer. This conclusion is in tune with the higher value of the polydispersity index. The mechanistic reason of this dilution effect is not known and is under investigation in our laboratory.

Table 10
Polymerization of the system 2-Methylpropene (1.2 M) /HN$_3$/TiCl$_4$ (5.6.10^{-2}M)/CH$_2$Cl$_2$, at -70°C, t = 30 min.

N°	[HN₃]/[TiCl₄]	Yield %	Mn$_{th}$** D	Mn$_{SEC}$* D	f$_{N3}$***	Used N₃ %
1	1.5	80	800	2500	0.4	10
2****	1.1	80	1100	2300	0.7	26
3	0.75	90	1600	2500	0.6	34
4	0.50	100	2400	3000	0.4	32

*In eq. p.St.
**Assuming one HN$_3$ per macromolecule (no transfer).
***determined by IR spectroscopy.
****This experiment when carried out in the temperature range -70°C -50°C give a polydispersity index of I = 1.35.

In order to show the field of application of the above conclusions, some informations must be given on a system in which initiation results from cocatalysis by HN$_3$.

Table 10 reports some experiments, and when the ratio $[HN_3]/[TiCl_4]$ is around 1, the polydispersity index is 1.35. Besides the azide content which is mentionned in Tables 9 and 10, all polymers contain tertiary chloride (as shown by ^{1}HNMR spectroscopy). It is clear that the same considerations as above can apply, i.e. that all macromolecules can grow up, except those which are terminated by an unsaturation. This means that terminal azide and terminal chloride are exchanging at equilibrium, and exchanging with the active species.

The conclusion drawn from these experiments is not to be considered as valid only in the case of 2-methylpropene. It was felt that this mechanism could be more general. For that reason it was decided to investigate a totally different polymerizing system, the 2-methyl-2-oxazoline in the presence of benzyl bromide in acetonitrile. It is well known that this is a living polymerization. The effect of the presence of TMSA was investigated at 80°C (15). Table 11 reports some experiments which allow to draw the main informations.

Table 11

Polymerization of the system 2-methyloxazoline/Benzyl bromide (BzBr = 0.044 M)/TMSA/ acetonitrile, at 80°C ([2-methyl-2-oxazoline] = 1.77 M).

N°	[TMSA] M	t h	Yield %	Mn_{the} *	Mn_{NMR} **	Mn_{SEC} ***	I_p ****	f_{N3} %
1	0.088	24	100	3575	3700	4100	1.3	89+5
2	0.132	24	100	3575	3000	4000	1.2	100+5
3*****	0.132	48	75	4710	3000	3800	1.3	100+5

*calculated from the monomer to BzBr ratio.
**Calculated from the NMR spectrum assuming one aromatic nucleus per chain.
***Determined by SEC (H_2O, light scattering)
****Polydispersity index.
***** Polymerization carried out according to the incremental monomer addition (IMA) technique, a second monomer charge (0.59 M) being introduced 24 hours after the first one (1.77 M).

Runs 1-2 clearly demonstrate that, despite the presence of TMSA and within experimental accuracy, the polymerization was living. It must be noticed that the experimental molecular weight were reasonably close to the theoretical one given by the ratio monomer to initiator concentrations. The polydispersity indeces were low, and as a whole, the behavior was the same as published some years ago by Saegusa et al. (16). The main point is that when the TMSA concentration was sufficiently high, the functionalization was apparently complete. However, run 3 indicates that when a new charge of monomer was introduced in such a system where the azide functionalization is complete, the polymerization of the second charge did not occur, while it does in the absence of TMSA. These observations are again in tune with our conclusion that terminal azide does not propagate but can exchange with the brominated chain ends. Thus, this scheme can explain the low polydispersity index. The reason why it was possible to obtain at the same a complete polymerization and a high level of functionalization is that the rate of polymerization is higher than that of functionalization. A complete analysis of this system is going to be published (15).

Similarly to the context of the organic azide, the polymerization of 2-methylpropene initiated by bis-1,4-(1-isothiocyano-1-methylethyl)benzene, called hereafter dicum-NCS, in the presence of $TiCl_4$ was tried at -70°C. One typical result is quoted on the next table 12.

Table 12

Polymerization of the system 2-methylpropene (IB)/dicumyl-NCS/$TiCl_4$/CH_2Cl_2, at -70°C.

N°	[IB] M	[dicum-NCS] M	[$TiCl_4$] M	Yield %	I***	f_{NCS}*	Mn_{SEC}** D
1	1.34	$1.5.10^{-2}$	$7.9.10^{-2}$	100	1.3	0.1	4900

*Determined by NMR spectroscopy (peak at 1.59 ppm= -CH_2-C-NCS).
**In eq. p.St.
***Polydispersity index.

It can be seen that in the conditions selected for this experiment, the polymerization yield is complete, but the functionalization by the isothiocyanate group was poor (10%) besides a very small amount of thiocyanate group. This last point shows that this experiment was carried out in such conditions that the thermodynamic equilibrium of exchange on chain ends was reached. NMR analysis revealed that the polymer contained an important fraction of tertiary chloride and no terminal unsaturation. This last observation is consistent with the low polydispersity index, since there was no transfer. Once again the equilibrium between chlorinated chain ends and their functionalized counterparts explains the low polydispersity index with a complete yield.

What can be learnt from this general survey ? It was first shown that functionalization can be observed with at least two pseudohalides, besides one experiment using CN published long time ago (17). Not only the synthesis of functionalized oligomers can be carried out in a one batch process, but also a control of the reaction can be obtained from adjustment of the conditions. This is particularly clear if the polymerization of 2-methylpropene is examined in the presence of an initiator or a pseudohalide donor such as TMSA.

An important observation was that a control of the molecular weight of PIB oligomers was obtained at sufficiently low temperature, i.e. at -70°C (using $TiCl_4$), and at 80°C for 2-methyl-2-oxazoline. These conditions are generally the ones which prevail obtaining controlled polymerization. Since in the case of $TiCl_4$ the chain ends were always containing at the same time tertiary chloride and a pseudohalide group, the low polydispersity index at complete yield can only be explained by an equilibrium involving both chain ends, according to the following scheme (mixed initiation or not):

Cocatalysis   Coinitiation   Direct Initiation

⇓

$$R-(-M-)_n-X + MtX'_n \Leftrightarrow R-(-M-)_{n-1}-M^+,C^- \Leftrightarrow R-(-M-)_nX' + MtXX'_{n-1} \text{ (A)}$$

⇓

polymer

where X is a pseudohalide, X' is a halide coming from the Lewis acid MtX'_n, C^- is a counteranion which contains a pseudohalide anion X^-, and $MtXX'_{n-1}$ is a Lewis acid in which a halide anion has been replaced by a pseudohalide anion. The interest of the equilibria (A) is that the polymer product is made uniform due to these exchange reactions which cannot be established with azided chain ends when the coinitiator is a too strong Lewis acid (BCl_3) or when multiple initiation can overcome this effect (BF_3).

It is important to note that a simple exchange of counteranion between different active species cannot account for I<2 for monofunctional growth, even if it is likely that the exchange is operating through the active species, as shown by equation (A) above, and not directly with the covalently bound pseudohalide X.

Now, the question is to know whether the chain end functionalized by the pseudohalide is able in itself to give access to a controlled polymerization without the presence of a chloride chain end. One system seems to indicate that it is possible to find conditions leading to controlled polymerization without any measurable amount of chlorinated chain ends. The main characteristics of this system were described somewhere else (18). In this paper, the polymerization of 2-methylpropene was studied initiated by BAMEB in the presence of $AlEt_2Cl$ in polar medium. It was shown that this system was reasonably living at -50°C or below, the number of macromolecules being equal to the number of initiator molecules (18). The next table 13 shows some experiments.

Table 13

Polymerization of 2-methylpropene in CH_2Cl_2 at -50°C initiated by the system BAMEB/AlEtCl$_2$

N°	[IB] M	[BAMEB] M	[AlEt$_2$Cl] M	Yield %	I**	f$_{N3}$***	Mn$_{SEC}$* D
1	0.053	$2.2.10^{-3}$	$1.7.10^{-2}$	98	1.3	2.0	1650
2	0.22	$2.2.10^{-3}$	$1.7.10^{-2}$	100	1.3	2.1	5800
3	0.84	$1.9.10^{-3}$	$1.5.10^{-2}$	98	1.3	2.1	24300

*In eq. p.St.
**Polydispersity index.
***determined by IR spectroscopy.

The functionalization of the macromolecules was perfect within experimental accuracy, and more particularly no chlorine atom was detected in conditions where the polymerization was controlled. It is then clear that in some conditions the presence of chlorinated termini is not a stringent requirement for obtaining a good molecular weight control. It was shown above that their presence participates to this control, while it is of course detrimental to the functionalization. However, it is then clear that the azide group can provide at the same time the dormant and the active species, providing specificity and selectivity. Most probably, the same situation could be met using the other pseudohalides, in some specific conditions which remain to be determined.

III- Conclusion

The use of additives containing a mobile pseudohalide function opens the way to functionalized oligomers in a one step synthesis from the monomer either through coinitiation or by exchange reaction. When the systems of 2-methylpropene polymerization described above were examined, it has been shown that the results varied progressively from systems where the functionalization was negligible or present in minor proportion (BCl_3, BF_3), to systems giving access at the same time to a fairly good functionalization in a one-batch process and to good molecular weight control ($AlEt_2Cl$). It is interesting to note that 2-methylpropene is not the only monomer susceptible to give functionalized oligomers by this technique.

IV- References

(1) H. A. Nguyen, H. Cheradame, E. Marechal, Makromol. Chem. **193**, 2495 (1992).

(2) R. Richter, H. Ulrich, in *The Chemistry of Cyanates and Their Thio derivatives*, Part 1, S. Patai Ed., John Wiley (New York) 1977, p. 619.

(3) W. H. Pearson, Wen-Kui Fang, J. Org. Chem. 1995, **60**, 4960-4961.

(4) H. Cheradame, J. Habimana, E. Rousset, F. Chen, Makromol. Chem. **192**, 2777-2789 (1991).

(5) H. Cheradame, J. Habimana, E. Rousset, F. Chen, Macromolecules **27**(3), 631 (1994).

(6) F. Duchemin, V. Bennevault-Celton, H. Cheradame, A. Macedo, Macromol. Chem. Phys. **199**, 2533-2539 (1998).

(7) Y. Peng, H. Dai, J. Liu, L. Cun, Chinese J. of Polym. Sci. **13**(4), 381 (1995).

(8) F. Duchemin, PhD thesis, Paris XII University, 27th march 1997.

(9) II. A. Nguycn, II. Chcradamc, Macromolcculcs 1995, **28**, 7942.

(10) A. G. Evans, G. W. Meadows, Trans. Faraday Soc. **46**, 327 (1950).

(11) P. A. S. Smith, D. W. Emerson, J. Am. Chem. Soc. **82**, 3076 (1960)

(12) C. Descours-Michallet, H. Cheradame, F. J. Chen, Macromolecules **26**, 1194 (1993).

(13) W. Buchman, B. Desmazieres, J. P. Morizur, H. A. Nguyen, H. Cheradame, Macromolecules 1998, **31**, 220-228.

(14) H. cheradame, P. Sigwalt, Bull. Soc. Chim. Fr. **1970**(3), 843.

(15) C. Guis, H. Cheradame, to be published.

(16). T. Saegusa, S. Kobayashi, A. Yamada, Makromol. Chem., **177**, 2271 (1976).

(17) H. Cheradame, J. Habimana, F. J. Chen, Makromol. Chem. **193**, 2647 (1992).

(18) B. Rajabalitabar, H. A. Nguyen, H. Cheradame, Macromolecules **29**(2), 514 (1996).

*Macromol. Symp. **157**, 137–142 (2000)*

Macromolecular Synthesis Using Mesoporous Zeolites

Keisuke Kageyama, Shau Mien Ng, Hiroyuki Ichikawa, and Takuzo Aida*

Department of Chemistry and Biotechnology, Graduate School of Engineering, The University of Tokyo, 7-3-1 Hongo, Bunkyo-ku, Tokyo 113-8656, Japan.

SUMMARY: Mesoporous zeolites such as MCM-41 were found to serve as nano-flasks for free radical polymerization of methyl methacrylate (MMA), where the formation of long-lived propagating radicals was observed. Al-MCM-41 with a Lewis-acidic aluminosilicate framework catalyzed living ring-opening polymerization of cyclic esters such as δ-valerolactone and ε-caprolactone, to give narrow molecular-weight-distribution polyesters. With Ti-MCM-41, a titanate-containing mesoporous silica, ring-opening polymerization of δ-valerolactone also took place to give a high molecular-weight polyester. On the other hand, with Ti-MCM-41 in the presence of methylaluminoxane (MAO), ethylene was polymerized to give a high molecular-weight, linear polyethylene.

Introduction

Mesoporous zeolites consist of a porous silicate framework with a regular hexagonal array of uniformly sized channels in nanoscopic diameter (15–100 Å), and are characterized by an exceptionally large surface area (~1000 m^{2}g^{-1}).[1] We were interested in utilization of mesoporous zeolites as inorganic nano-flasks for controlled polymerizations, as they could serve as restricted but adequately large spaces for the synthesis of macromolecular compounds. The present paper reports some results of our recent study on utilization of mesoporous zeolites for (1) free radical polymerization of vinylic monomers, (2)

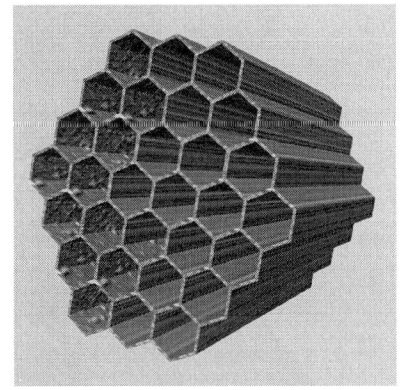

Scheme 1: Schematic structure of mesoporous zeolite (MCM-41).

ring-opening polymerization of lactones, and (3) coordination polymerization of ethylene.

        CCC 1022-1360/00/$ 17.50+.50/0

Free Radical Polymerization of Methyl Methacrylate

In free radical polymerization, the propagation reaction is generally more difficult to control than in ionic polymerizations, due to the irreversible termination of the growing polymer radicals through recombination and disproportionation. In order to control free radical polymerization, elegant methods by utilization of transition metals for the stabilization of growing polymer radicals have been developed.[2] On the other hand, in addition to such "chemical approaches", so-called "physical approaches" have also been investigated by using some confined spaces such as organic micelles[3] and inclusion crystals for spatial isolation of each growing polymer radical. We have been exploring the potential of a mesoporous zeolite such as MCM-41 as a nano-flask for free radical polymerization of methyl methacrylate.[4]

The free radical polymerization of methyl methacrylate (MMA) within the mesoscopic channels of MCM-41 at 100 °C proceeded to give a high molecular-weight polymer under appropriate conditions. Typically, a mixture of MMA, 2,2'-azoisobutyronitrile (AIBN) (mole ratio MMA/AIBN of 200), and thermally degassed MCM-41 powder was subjected to several freeze-to-thaw cycles. Then, the resulting "guest-including MCM-41 powder" was isolated by filtration from the suspension, and put into a polymerization tube. The tube was heated at 100 °C under N_2 to initiate the polymerization. After 2 h, the contents of the tube were transferred into tetrahydrofuran containing hydroquinone as a radical scavenger, and the resulting suspension was refluxed for 3 days to extract polymeric fractions from the channels.

The polymer obtained under the above conditions (71 % yield based on the amount of MMA) had a high molecular weight (Mn = 360,000, Mw/Mn = 1.7), as observed by gel permeation chromatography (GPC). On the other hand, without MCM-41 under otherwise identical conditions to the above, a polymer with a much lower molecular weight (Mn = 36,000, Mw/Mn = 2.8) was obtained. These results suggest a

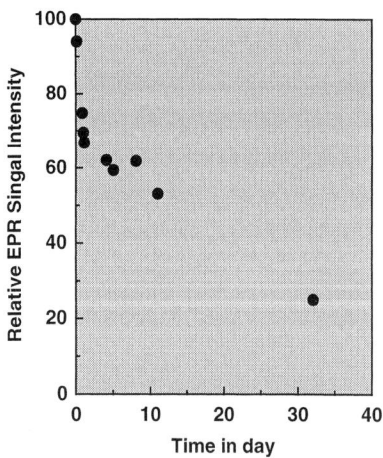

Fig. 1: Polymerization of MMA initiated with 2,2'-azoisobutyronitrile (AIBN) at a mole ratio $[MMA]_0/[AIBN]_0$ of 200 within the MCM-41 channels. Change in intensity of the EPR signal of the polymerization mixture after heating at 100 °C for 2 h.

possible elongation of the propagating polymer radical within the mesoporous channel. Accordingly, electron paramagnetic resonance (EPR) spectroscopy showed a signal due to the propagating polymer radical, which decayed slowly as shown in Fig. 1, where about 30% of the initial intensity remained even after a month.

Taking the above results into account, we explored the possibility of molecular weight control by changing the monomer-to-initiator mole ratio within the MCM-41 channels. The polymerizations of MMA within the MCM-41 channels initiated by benzoyl peroxide (BPO) at three different mole ratios of 100, 150, and 200 proceeded to attain about 25% monomer conversion in 6 days. The molecular weights of the resulting PMMAs showed a linear correlation to the mole ratio $[MMA]_0/[BPO]_0$ (Fig. 2).

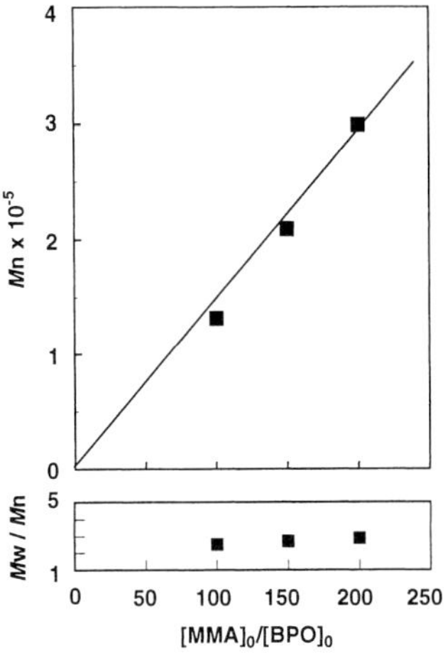

These observations indicate a potential utility of the confined channels of MCM-41 for controlled free radical polymerization.

Fig. 2: Polymerization of MMA initiated with benzoyl peroxide (BPO) within the MCM-41 channels. Mn (■) (Mw/Mn (●)) $-[MMA]_0/[BPO]_0$ relationship.

Ring-Opening Polymerization of Lactones

Polyesters from lactones are important precursors for several polymeric materials and also have attracted attention as biodegradable polymers. We have reported that ring-opening polymerization of lactones proceeds with bulky Lewis acids in the presence of alcohols in a living manner.[5] Along this line, mesoporous zeolites with a Lewis acidic surface are expected to serve as new class of inorganic polymerization catalysts. In fact, Al-MCM-41, which consists of Lewis acidic aluminosilicate framework, catalyzed controlled ring-opening polymerization of lactones such as δ-valerolactone (VL) and ε-caprolactone (CL) initiated with alcohols, to give narrow molecular-weight-distribution (MWD) polyesters.[6]

For example, when a mixture of VL and BuOH at a mole ratio of 11 was added under N_2 to a flask containing Al-MCM-41 (Si/Al = 17, 0.1 g), and the resulting suspension was stirred magnetically at 50 °C, the polymerization proceeded to attain 96.4% monomer conversion in 237 h, and gave a narrow MWD polymer with Mn of 1,200 (Mw/Mn = 1.07). The molecular weight of the polymer could be controlled over a wide range by changing the feed molar ratio of the monomer to the alcohol, while the MWD stayed in a narrow range (Fig. 3).

To investigate the living character of the polymerization, a sequential two-stage polymerization of VL and CL was attempted. The first-stage polymerization of VL with Al-MCM-41/BuOH was carried out at 50 °C to produce a prepolymer (Mn = 1,200, Mw/Mn = 1.07). After the complete consumption of VL, CL was added at 100 °C to the system, then the second-stage polymerization of CL took place to attain 100% monomer conversion. The GPC profile of the polymerization mixture showed a unimodal, sharp chromatogram at a higher molecular weight region (Mn = 3,200, Mw/Mn = 1.20), indicating the formation of a block copolymer of VL and CL. Thus, the polymerization of lactones with the Al-MCM-41/ROH has a living character. In contrast to Al-MCM-41, a microporous zeolite

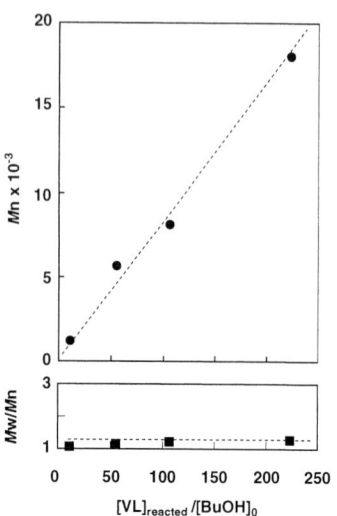

Fig. 3: Polymerization of δ-valerolactone (VL) initiated with BuOH in the presence of Al-MCM-41. Mn (●) (Mw/Mn (■)) –[VL]$_{reacted}$/[BuOH]$_0$ relationship.

(zeolite-Y) with much narrower pores (pore diameter = 8 Å) was not effective for the polymerization of VL at 50 °C. This result indicates an importance of the large surface area and a wide pore width of Al-MCM-41 for the accessibility of the monomer and the polymer to the propagating site.

For the mechanism of the polymerization, we propose the "activated monomer mechanism" (Scheme 2), where the monomer is activated through a cooperative interaction with the Lewis acidic aluminum atom and the Brønsted acidic SiOH functionality, since neither pure silicate MCM-41 nor Al–MCM-41 having masked SiOH functionalities with methyl groups was effective for the polymerization.

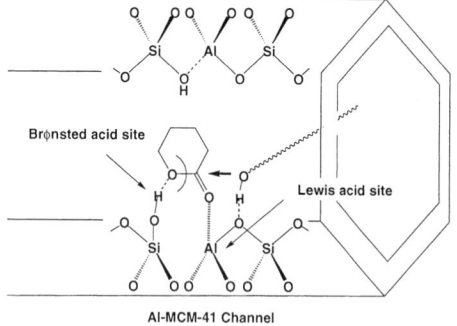

Scheme 2: Proposed mechanism for the polymerization of δ-valerolactone (VL) with alcohol within the aluminosilicate channels of Al-MCM-41.

The above results prompted us to investigate the polymerization of VL with MCM-41 containing other Lewis-acidic metal ions. Thus, a titanate-containing MCM-41 (Ti-MCM-41), prepared by mounting Cp_2TiCl_2 on the MCM-41 surface, followed by calcination (Scheme 3),[7] was chosen. Interestingly, the polymerization profile was much different from that with Al-MCM-41.[8] For example, when a mixture of VL (4 mL) and Ti-MCM-41 (0.25 g) was heated at 50 °C under N_2 without solvent, the polymerization took place very slowly to attain 12% monomer conversion in 43 h. It should be noted that the polymerization mixture turned into a gel even at such a low monomer conversion. When the gel was poured into THF at room temperature, a small amount of insoluble polymeric materials formed and precipitated. GPC analysis of the THF-insoluble fraction with chloroform as eluent showed a multimodal

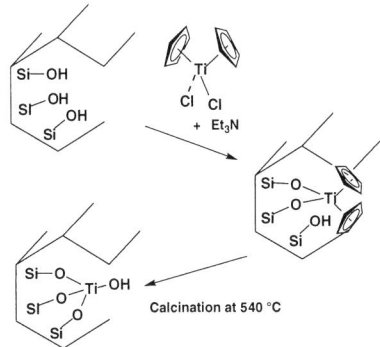

Scheme 3: Method for the preparation of Ti-MCM-41.

MWD, suggesting the presence of multiple active sites in Ti-MCM-41. However, to our surprise, the polymer contained an ultrahigh-molecular-weight fraction with a peak-top molecular weight higher than 10^6. Fine-tuning of the polymerization with Ti-MCM-41 may be the subject worthy of further investigation.

Polymerization of Ethylene

Polymerization of olefins by supported catalysts have been of interest from an industrial point of view. We have found that Ti-MCM-41 in the presence of methylaluminoxane (MAO) initiates polymerization of ethylene, to give a high molecular-weight, linear polyethylene.

Typical polymerization procedure is as follows: A toluene solution of MAO (Al/Ti = 100) was added to a glass autoclave (100 mL) containing dried Ti-MCM-41 powder (0.05 g). To this autoclave was introduced ethylene gas at an initial pressure of 10 atm with stirring at room temperature. The pressure was observed to drop gradually as the polymerization proceeded, where a complete consumption of ethylene was attained in 70 h. Then, a small amount of methanolic HCl was added to the polymerization mixture, and an insoluble fraction was isolated by filtration and washed with methanol. After extraction with o-dichlorobenzene at 180 °C, the polymeric fraction, free from inorganic compounds, was subjected to GPC to determine the molecular weight. Under appropriate conditions, a polyethylene with a very high molecular weight such as 1,500,000 was obtained. This is in sharp contrast with the case under homogeneous conditions using the Cp_2TiCl_2/MAO system, where the molecular weight of the polymer was only as low as 100,000. Using the Ti-MCM-41/MAO system, a polyethylene with a much higher molecular weight was obtained by reducing the mole ratio Al/Ti. In these cases, ^{13}C NMR spectroscopy showed that the produced polymers are linear without any branch structures. It should be also noted that $AlMe_3$ and $AlEt_3$ serve as effective co-catalysts for the polymerization of ethylene with Ti-MCM-41. Thus, Ti-MCM-41 has a potential utility for the synthesis of linear, high molecular-weight polyethylene.

Conclusion

A recent progress in the chemistry of mesoporous zeolites has made it possible to modify chemical and physical properties of the channels. In the present paper, we have shown that mesoporous zeolites are interesting materials as nano-flasks for controlled polymerization of a variety of monomers. Fine-tuning of the surface properties by incorporation of transition metal atoms onto/into the silicate framework is one of the subjects worthy of further investigation.

Reference

1. J. S. Beck, J. C. Vartuli, W. J. Roth, M. E. Leonowicz, C. T. Kresge, K. D. Schmitt, C. T. -W. Chu, D. H. Olson, E. W. Sheppard, S. B. McCullen, J. B. Higgins, J. L. Schlenker, *J. Am. Chem. Soc.*, **114**, 10834 (1992).
2. (a) M. Kato, M. Kamigaito, M. Sawamoto, T. Higashgimura, *Macromolecules*, **28**, 1721 (1995). (b) J. –S. Wang, K. Matyaszewski, *J. Am. Chem. Soc.*, **117**, 5614 (1995).
3. K. Horie, D. Mikulásová, *Makromol. Chem.*, **175**, 2091 (1974).
4. S. M. Ng, S. Ogino, T. Aida, K. A. Koyano, T. Tatsumi, *Macromol. Rapid Commun.*, **18**, 991 (1997).
5. M. Akatsuka, T. Aida, S. Inoue, *Macromolecules*, **28**, 1320 (1995).
6. K. Kageyama, S. Ogino, T. Tatsumi, T. Aida, *Macromolecules*, **31**, 4069 (1998).
7. T. Maschmeyer, F. Rey, G. Sankar, J. M. Thomas, *Nature*, **378**, 159 (1995).
8. K. Kageyama, T. Tatsumi, T. Aida, *Polymer J.*, **31**, 1005 (1999).

Synthesis of Novel Hydrocarbon Polymers Containing 6-Membered Rings in the Main Chain : Living Anionic Polymerization of 1,3-Cyclohexadiene

Itaru Natori*, Kimio Imaizumi

Corporate Research and Development Administration (Europe), Asahi Chemical Industry, London Office, Gainsborough House Suite 607, 81 Oxford Street, London W1R 1RB, U.K.

SUMMARY:The *n*-butyllithium (n-BuLi) / *N,N,N',N'*-tetrametylethylene-diamine (TMEDA) system (the molar ratio of TMEDA to *n*-BuLi higher than 4/4) has been found to polymerize 1,3-cyclohexadiene (1,3-CHD) to produce "living" polymer having narrow molecular weight distribution with well-controlled polymer chain length. Binary and ternary block copolymers with narrow molecular weight distribution could be synthesized from 1,3-cyclohexadiene, styrene, and butadiene with very high efficiency. These polymers and their hydrogenated derivatives have excellent thermal, mechanical, chemical, and optical properties for the new industrial materials.

Introduction

The synthesis of hydrocarbon polymers containing alicyclic structure in the main chain is one of the most interesting subjects in basic and practical aspects, since a dramatic improvement of thermal stability, chemical stability, and mechanical strength is expected for such polymers. In particular, directly connected cyclohexane rings are selected as the most favorable compound out of all alicylic monomeric unit as target of polymer chain for ideal hydrocarbon polymers. Thus, the most important problem to be solved was the method of direct connection of cyclohexane rings in the main chain. For example, the polymerization of 1,3-cyclohexadienen (1,3-CHD) expected to provide a preferable prepolymer of poly(cyclohexane) which has directly connected cyclohexene rings. If well-controlled polymerization of 1,3-CHD will be possible, directly connected cyclohexane rings with well-defined structure can be prepared by the hydrogenation of poly(1,3-cyclohexadiene) (Figure 1).

---

*e-mail : a8710317@ut.asahi-kasei.co.jp

© WILEY-VCH Verlag GmbH, D-69469 Weinheim, 2000   CCC 1022-1360/00/$ 17.50+.50/0

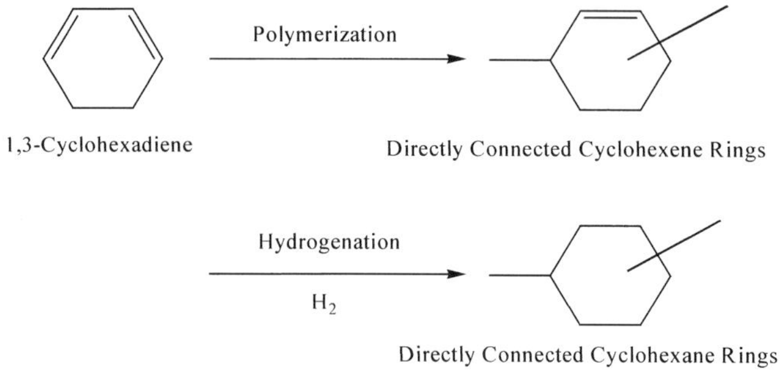

Polymerization

1,3-Cyclohexadiene

Directly Connected Cyclohexene Rings

Hydrogenation

H_2

Directly Connected Cyclohexane Rings

Figure 1. Synthesis of hydrocarbon polymers with directly connected cyclohexane rings. Polymerization of 1,3-cyclohexadiene and hydrogenation of poly(1,3-cyclohexadiene).

However, the polymerization of 1,3-CHD has been reported to be difficult in the studies under various conditions including the case of anionic polymerization with alkyllithium as initiator. The obtained polymers were of low molecular weight or in low yield [1]. In our investigation, the new polymerization technology for 1,3-CHD was examined by using various polymerization methods. Finally, a specific initiator system for living anionic polymerization of 1,3-CHD was found for the first time [2].

In this paper, we report the first successful example of living anionic polymerization of 1,3-CHD, and then the block copolymerization with styrene (St) and butadiene (Bd). Furthermore, the microstructure and properties of 1,3-CHD homopolymer, block copolymers, and their hydrogenated derivatives are also described.

Living Anionic Polymerization of 1,3-Cyclohexadiene

In the previous paper, we reported the first successful example of living anionic polymerization of 1,3-cyclohexadiene with n-BuLi/TMEDA system as initiator[2a]. The molar ratio of n-BuLi and TMEDA composing the initiator system strongly influenced the anionic polymerization of 1,3-CHD [2d]. The polymerization attained the living nature as the ratio of TMEDA increased. When the ratio of TMEDA to n-BuLi was higher than 4/4, obtained polymers had very narrow molecular weight distribution. On the other hand, no polymeric product was obtained in the polymerization initiated by n-BuLi in the absence of TMEDA($[1,3\text{-CHD}]_0/[\text{Li}]_0=125$, 120 min, 40°C). These results suggest that the propagation ends of living poly(1,3-cyclohexadiene) is composed of 1/1 molar ratio of Li and TMEDA (Table 1) [2d].

Table 1. Polymerization of 1,3-cyclohexadiene by the *n*-BuLi/TMEDA Systems [a]

No.	Iinitiator system	Yield (%)	Mn	Mcalc.[b]	Mw/Mn
1	n-BuLi	0			
2	n-BuLi/TMEDA(4/2)	72	16,000	7,200	1.52
3	n-BuLi/TMEDA(4/3)	100	11,100	10,000	1.12
4	n-BuLi/TMEDA(4/4)	100	10,600	10,000	1.07
5	n-BuLi/TMEDA(4/5)	100	11,600	10,000	1.06
6	n-BuLi/TMEDA(4/6)	100	10,800	10,000	1.08
7	n-BuLi/TMEDA(4/7)	100	10,100	10,000	1.06
8	n-BuLi/TMEDA(4/8)	100	11,300	10,000	1.07

[a] $[1,3\text{-CHD}]_0/[\text{Li}]_0 = 125$. [Monomer(s)]/[Solvent]=10/90. Polymerization was carried out in cyclohexane. Reaction temp. 40℃. Reaction time 120min. Mn and Mw/Mn were estimated by GPC, using polystyrene as standards. [b] Mcalc: $FW(C_6H_8) \times 125 \times \text{Yield(\%)} \times 10^{-2} = 10,000 \times \text{Yield(\%)}$.

The characteristics of living anionic polymerization of 1,3-CHD initiated by some *n*-BuLi/TMEDA systems (4/5), (4/2), and (4/0.5) in cyclohexane was examined in further detail ($[1,3\text{-CHD}]_0/[\text{Li}]_0 = 250$, 240 min, 40℃).

The results are shown in Figures 2 and 3.

Figure 2

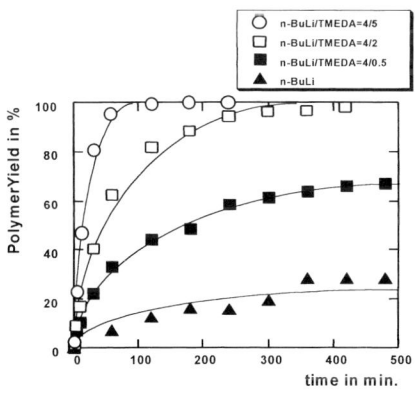

Figure 3

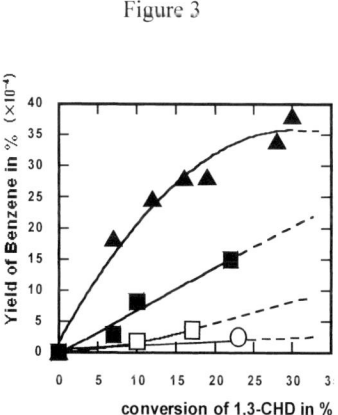

Figure 2. Conversion-time relationship of the polymerization of 1,3-CHD with some *n*-BuLi/TMEDA systems in cyclohexane at 40℃. [1,3-CHD]$_0$/[Li]$_0$=250, the ratio of *n*-BuLi/TMEDA(4/5), (4/2) and (4/0.5).

Figure 3. The relationship between the conversion of 1,3-CHD and the amount of benzene fromed in the anionic polymerization of 1,3-CHD initiated by some *n*-BuLi/TMEDA systems and *n*-BuLi in cyclohexane at 40℃. [1,3-CHD]$_0$/[Li]$_0$=250, the ratio of *n*-BuLi/TMEDA(4/5), (4/2), and (4/0.5).

Figure 2 clearly shows that the rate of polymerization and polymer yield increased with the ratio of TMEDA to *n*-BuLi. On the other hand, the polymer yield was considerably low in the absence of TMEDA and the increase in polymer yield was not observed after 360 min. The molecular weight distribution of obtained polymers became narrower with the ratio of TMEDA to *n*-BuLi (*n*-BuLi/TMEDA (4/5) : Mn=18,700, Mw/Mn =1.09, conversion 100%; (4/2) : Mn=28,200, Mw/Mn=1.41, conversion 100%; (4/0.5) : Mn=16,300, Mw/Mn=2.42, conversion 77%; *n*-BuLi : Mn=3,500, Mw/Mn=2.27, conversion 30%). The yield of benzene was found to decrease with the ratio of TMEDA to *n*-BuLi (Figure 3). The formation of benzene is considered to be the results of the decomposition of cyclohexadienyllithium (CHDLi) with hydride elimination and the abstraction of the allylic hydrogen of 1,3-CHD by CHDLi, respectively[1]. Obviously, in the polymerization of 1,3-CHD with the *n*-BuLi/TMEDA (4/5) system, there is neither transfer reaction nor termination reaction by the abstraction of allylic hydrogen of 1,3-CHD by organolithium species.

Block Copolymerization of 1,3-Cyclohexadiene and Styrene [2b]

The living prepolymer of 1,3-CHD was prepared by the polymerization with the *n*-BuLi/TMEDA (4/5) system ([1,3-CHD]$_0$ /[Li]$_0$=125, 120min, 40℃, 100% conversion). Styrene (St) was then added into the above reaction mixture ([St]$_0$/[Li]$_0$=384). After 120 min, the conversion of St was 100%. 1,3-CHD was added into the above binary block copolymer solution ([1,3-CHD]$_0$/ [Li]$_0$=125) to continue the reaction. The conversion of 1,3-CHD after 300 min was 96%. The number average molecular weight (Mn) and molecular weight distribution (MWD) of this ternary block copolymer were 64,000 and Mw/Mn=1.14, respectively (Figure 4). The obtained 1,3-CHD-St-1,3-CHD ternary block copolymer was soluble in cyclohexane, tetrahydrofuran (THF), and chloroform. A transparent colorless strong plastic film could be obtained by casting from the cyclohexane solution of this polymer.

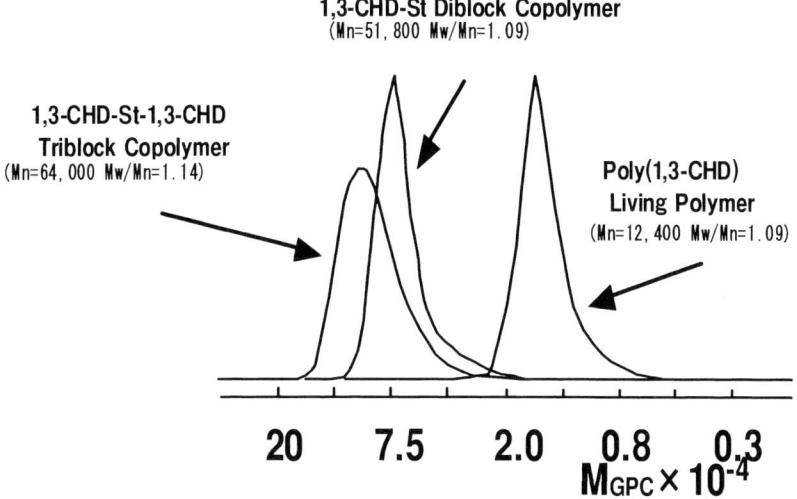

Figure 4. Block copolymerization of 1,3-cyclohexadiene and styrene.

Block Copolymerization of 1,3-Cyclohexadiene and Butadiene [2b]

The living prepolymer of 1,3-CHD was prepared by the polymerization with the n-BuLi/TMEDA (4/5) system ([1,3-CHD]$_0$/[Li]$_0$=125, 120min, 40°C, 96% conversion). Butadiene (Bd) was then added into the above reaction mixture ([Bd]$_0$/[Li]$_0$=370). After 60 min, the conversion of butadiene was 99%. 1,3-CHD was added into the above binary block copolymer solution ([1,3-CHD]$_0$/[Li]$_0$=125) to continue the reaction. The conversion of 1,3-CHD after 300 min was 95%. The number average molecular weight (Mn) and molecular weight distribution (MWD) of this ternary block copolymer were 41,000 and Mw/Mn=1.15, respectively. The obtained 1,3-CHD-Bd-1,3-CHD ternary block copolymer was soluble in cyclohexane, tetrahydrofuran (THF), and chloroform. A transparent colorless tough elastic film could be obtained by casting from the cyclohexane solution of this polymer.

Microstructure of Poly(1,3-cyclohexadiene) [2d]

Poly(1,3-cyclohexadiene) has a structure of the main chain consisting of units formed by 1,2-addition (1,2-unit) and 1,4-addition (1,4-unit) (Figure4). The ratio of 1,2-unit and 1,4 -unit can be determined by 2D-NMR(H-H COSY method) [2d]. As for the polymer initiated by the *n*-BuLi/TMEDA systems, the content of the 1,2-units in the polymer chain increased with the ratio of TMEDA to *n*-BuLi. When the ratio of TMEDA to *n*-BuLi was higher than 4/4, the content of the 1,2-units was almost the same (51∼54%).

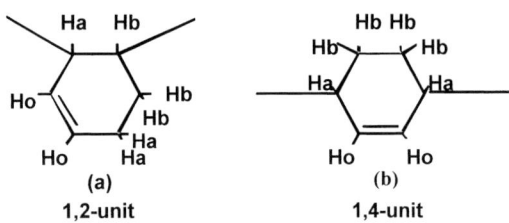

Figure 5. Structural formulas of 1,2-unit (a) and 1,4-unit (b).

Thermal, Mechanical, Chemical, and Optical Properties of Poly(1,3-cyclohexadiene) (PCHD) and Hydrogenated Poly(1,3-cyclohexadiene) [Poly(cyclohexane) :PCHE] [2c]

As described above, the novel hydrocarbon polymers containing 6-menbered rings directly connected in the main chain can be synthesized by living anionic polymerization of 1,3-cyclohexadiene. Various types of 1,3-CHD/Bd, 1,3-CHD/St, and 1,3-CHD-isoprene(Ip) block copolymers can be also synthesized by living anionic polymerization of 1,3-cyclohexadiene and styrene or butadiene (or isoprene).

As shown in Table 2, completely hydrogenated poly(1,3-cyclohexadiene) [poly-(cyclohexane) :PCHE] is composed of cyclohexane rings directly connected in the main chain, and has the highest glass transition temperature (Tg) among all hydrocarbon polymers. PCHE also has some excellent properties such as low specific gravity, high heat resistance, high flexural modulus, high transparency, low dielectric constant, and low water absorption. The refractive index is almost same as glass (SiO_2).

Table 2.   Properties of poly(1,3-cyclohexadiene) (PCHD) and hydrogenated
poly(1,3-cyclohexadiene)[poly(cyclohexane) : PCHE]

	PCHD	PCHE	PMMA	PC
Specific Gravity	1.03	0.99	1.19	1.19
Glass Transition Temperature (°C)	175	238	100	150
Heat Distortion Temperature (°C)	150	183	90	140
Flexural Modulus (GPa)	4.0	6.1	3.4	2.4
Transparency (%)	92	91	93	90
Reflective Index	1.43	1.51	1.58	1.49
Dielectric Constant (ε)	2.5	2.3	4.0	3.0
Water Absorption (%)	<0.01	<0.01	0.3	0.3

[a] PMMA : Poly(methyl methacrylate), PC : Polycarbonate

As a special optical property of PCHE, because there is no functional group in the
polymer chain of PCHE, it has an outstanding wavelength dependence for optical uses
among all other amorphous polymers (Figure 6).

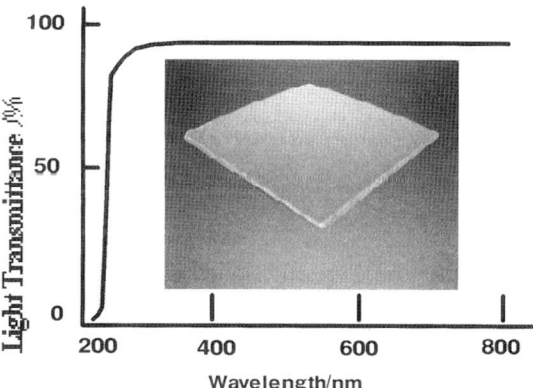

Figure 6. Photograph and wavelength dependence of PCHE.

Completely hydrogenated poly(1,3-cyclohexadiene)s such as hydrogenated 1,3-CHD-
Bd-1,3-CHD triblock copolymer has good weatherability for outdoor uses (Figure 7).
This result suggests that poly(cyclohexane) structure has good UV resistance.

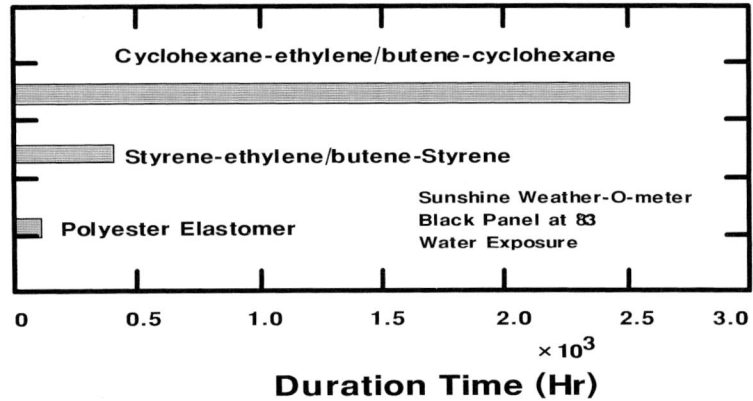

Figure 7. Weatherability of hydrogenated 1,3-cyclohexadiene-butadiene-1,3-cyclohexadiene triblock copolymer (PCHE-HBd-PCHE)

Conclusion

The *n*-BuLi/TMEDA system is an excellent initiator for the living anionic polymerization of 1,3-cyclohexadiene. The hydrocarbon polymers containing directly connected 6-membered rings in the main chain can be synthesized effectively by using this new technology. These novel hydrocarbon polymers and block copolymers have excellent properties to develop many high-performance industrial materials.

References

1. (a) G. Lefebver, F. Dawans, *J. Polym. Sci., Part A* **2**, 3277(1964). (b) P. E. Cassidy, C. S. Marvel, *J. Polym. Sci., Part A* **3**, 1533(1965). (c) H. Lussi, J. Barman, *J. Helv. Chim. Acta*, **50**, 1233(1967). (d) L. A. Mango, R. W. Lenz, *Polym. Prepr.(Amer. Chem. Soc., Div. Polym. Chem.)*, **12**, 402(1971). (e) L. A. Mango, R. W. Lenz, *Makromol. Chem.*, **163**, 13(1973). (f) X. F. Zhong, B. Francois, *Makromol. Chem.*, **191**, 2735(1990). (g) B. Francois, X. F. Zhong, *Makromol. Chem.*, **191**, 2743(1990).

2. (a) I. Natori, *Macromolecules*, **30**, 3696(1997). (b) I. Natori, S. Inoue, *Macromolecules*, **31**, 982(1998). (c) I. Natori, K. Imaizumi, H. Yamagishi, M. Kazunori, *J. Polym. Sci., Part B, Polym. Phys.*, **36(10)**, 1657(1998). (d) I. Natori, S. Inoue, *Macromolecules*, **31**, 4687(1998).

Towards the Control of the Reactivity in High Temperature Anionic polymerization of Styrene : Retarded Anionic Polymerization

3 – Influence of triisobutylaluminum on the reactivity of polystyryllithium species.

Philippe Desbois[1,2], Michel Fontanille*[,1], Alain Deffieux[1], Volker Warzelhan[2], Christian Schade[2]

[1] Laboratoire de Chimie des Polymères Organiques, UMR ENSCPB - Université Bordeaux-1 - CNRS, BP 108, 33402 Talence-CEDEX, France
[2] BASF Aktiengesellschaft, Polymer Laboratory, 67056 Ludwigshafen, Germany

SUMMARY: the addition of triisobutylaluminum to a lithiated-based anionic polymerization of styrene leads to the formation of aluminate complexes and provokes a drastic reduction of the styrene polymerizability in relation with the Al/Li ratio used. The reaction is first order in monomer and active species and is stopped for $iBu_3Al/PSLi > 1$. From U.V. spectrometry, kinetic studies and viscometric measurements, it was possible to suggest the presence of various mixed complexes of different stoichiometries. The strong increase in the thermal stability of the new propagating species together with their much lower reactivity permits the controlled bulk polymerization of styrene at high temperature.

Introduction

For long time and until a recent period, polymer chemists tried to increase the reactivity of active centers propagating chain polymerizations in order to obtain a better productivity of the corresponding processes. This trend has found limitations where industrial chemists encountered severe heat transfer problems. In such instances, they are trying to improve the control of the processes by reducing the overall reactivity.

In previous papers[1-3] it was shown that the anionic polymerization of styrene initiated by an alkyllithium in hydrocarbon solution can be strongly retarded by addition of dialkylmagnesium to the system. The reduction of the polymerizability is due to the formation of "ate" complexes between magnesium and lithium species, which provokes a strong decrease of the reactivity of propagating centers.

Thus, it is possible to perform anionic polymerizations at temperatures higher than 100°C while keeping the process under control. The persistence of propagating centers is strongly increased simultaneously to the reactivity reduction, even though the ratio k_p/k_t decreases slightly with raising temperatures. The bulk retarded anionic polymerization of styrene thus becomes possible at temperatures higher than T_g and an important new process for the production of polystyrene could be derived from Li/Mg-based systems as initiators.

In the light of the present knowledges and techniques and with the same objectives, it was tempting to revisit the system described by Welch[4] in 1960. This author had shown that addition of small amounts of aluminum alkyls to lithium alkyls ([Al]/[Li] < 1) leads to systems able to polymerize styrene but with a rate of polymerization lower than that observed with alkyllithium alone. Welch was considering that the decrease of the overall polymerization rate was due to the formation of a 1:1 RLi/AlR_3 inactive complex which was lowering the concentration of the reactive "free" RLi.

A rate reduction of the alkyllithium initiated polymerizations by addition of Al derivatives has been obtained and studied comprehensively in the case of polar monomers like methacrylates[5-8]. The presence of Al derivatives yields better kinetic control and (in addition, in several cases) a stereoregulation of the propagation process.

The present study was performed in order to demonstrate that mixed lithium/aluminum-based systems could be useful initiators for styrene anionic polymerization performed at relatively high temperatures. The main interest is to establish that the corresponding initiating systems preserve a living character for the polymerizations as for those performed at much lower temperatures.

In the aim to perform the polymerization in experimental conditions leading to a behavior close to that observed in bulk, experiments were realized in cyclohexane solution. Indeed, this solvent plays only the role of diluent and does not interfere on the intrinsic reactivity of the propagating centers. In such conditions the established data could be easily adapted to the bulk polymerization.

Results and Discussion

a) U.V.-Visible Study of Active Species

Addition of triisobutylaluminium (i-Bu$_3$Al) to polystyryllithium (PSLi) seeds in cyclohexane solution (λ_{max} = 326 nm) leads to the formation of a new peak located at 280-288 nm depending on the value of r = [Al]/[Li]. Simultaneously, the signal of PSLi decreases and disappears totally for r = 1 (Fig.1). For higher values of r, the U.V-Vis. absorption spectrum does not change and such a behavior is quite different of that observed with magnesiate complexes[1,2]. The modification of the U.V. spectrum is consistent with the formation of a 1/1 complex as suggested by Welch[4].

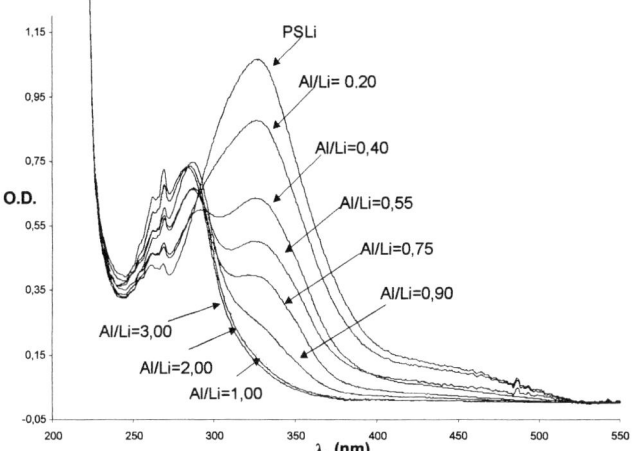

Fig. 1: Effect of addition of increasing amounts of i-Bu$_3$Al on U.V. spectrum of PSLi in cyclohexane ([PSLi] = 5.10^{-3} mol/L).

It is important to point out that for ratio close to 1 (Al/Li=0,8-1) the formation of the final spectrum needs several hours and proceeds in two stages (Fig. 2). The existence of an isosbestic point on the recorded spectra suggests the instantaneous formation of a "primary" complex between lithiated species and i-Bu$_3$Al (the "kinetic" complex) which rearranges into aλ more stable one (the "thermodynamic" complex).

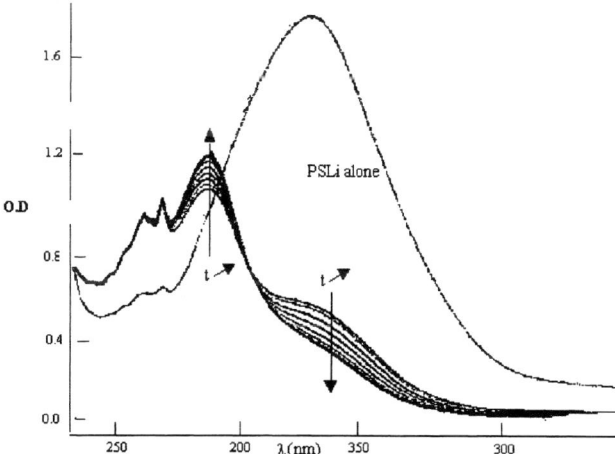

Fig. 2: Effect of aging on the U.V. spectrum of PSLi/i-Bu$_3$Al system. Interval of time between 2 spectra: 2 hours (r = 0.9 and T = 50°C).

Thus, the following steps and equilibria could be written:

2 PSLi

$$(PSLi)_2 + 2\ AlR_3 \rightleftharpoons \frac{x}{n}[PSLi,AlR_3]_n + [(2-x)PSLi,(2r-x)AlR_3] \quad \text{and}$$

$$\frac{x}{n}[PSLi,AlR_3]_n + [(2-x)PSLi,(2r-x)AlR_3] \rightleftharpoons \frac{y}{n}[PSLi,AlR_3]_n + [(2-y)PSLi,(2r-y)AlR_3]$$

$$\qquad\qquad\qquad\qquad\qquad\qquad\qquad\qquad (a) \qquad\qquad\qquad (b')$$

The second equilibrium is reached after a period of time of about 15 hours at 50°C.

b) Kinetic Studies

Table 1. Kinetic data for styrene polymerizations performed at 50°C in cyclohexane solution and initiated with Li/Al-based systems; $r = [Al]/[Li.]$. (a)

$\dfrac{[Bu_3Al]}{[RLi]}$	[PSLi] mol.L^{-1}	Rp/[M] min^{-1}	k_{app} mol$^{-0.5}$L$^{0.5}$min^{-1}	$\bar{M}_{n,th(Li)}$	$\bar{M}_{n,GPC}$	$I_p=\bar{M}_w/\bar{M}_n$
0	$6.2.10^{-3}$	0.13	1.6	10200	10900	1.05
0,50	$6.5.10^{-3}$	0.044	0.54	7250	12000	1.23
0,60	$6.3.10^{-3}$	0.021	0.26	7450	11000	1.28
0,75	$7.7.10^{-3}$	$7.0.10^{-3}$	0.08	6600	6700	1.30
0,95	$4.8.10^{-3}$	$3.1.10^{-4}$	0.004	13100	13600	1.08
1	$5.1.10^{-3}$	0	0	-	-	-

Data reported in Table 1 show that, whatever the values of r, the experimental number average molar masses are in satisfactory agreement with those calculated from initial [RLi] which exhibit the non-participation of i-Bu$_3$Al as chain carrier. Moreover, the polydispersity indexes I_p (Table 1) and the linearity of $\log_e[M_0]/[M]$ vs time (Fig. 3) indicate the absence of significant termination reactions. Similar results have been observed at higher temperature (T = 100°C).

The kinetic order with respect to active centers was calculated from these data. It equals to 0.98 and thus is considered as first order. This means that the proportion of actually active species is not affected by their concentration even if they are in equilibrium with other species with a different reactivity. The whole of this behavior is consistent with a living character of the process; thus, assuming the concentration of active species equal to that of RLi it is

possible to deduce an apparent rate constant of propagation : $\qquad k_{app} = \dfrac{R_P}{[S]\,[PSLi]}$

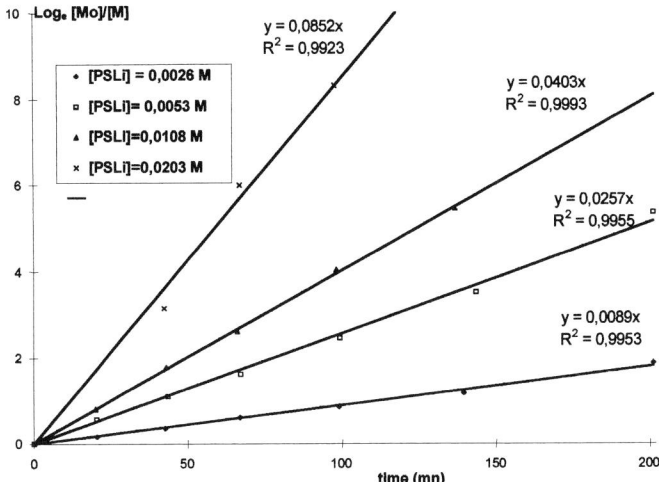

Fig. 3: Variation of $\log_e[M_0]/[M]$ *vs* time for different values of [Li]. r = 0.80; T = 100°C; cyclohexane.

At 50°C, k_{app} for r = 0.50 is 3 times lower than that measured for r = 0. With r = 0.95, the relative rate is further reduced to 1/400. Note that even for this ratio the number of PS chains formed is consistent with the contribution of the initial RLi species. The plot which represents the variation of $\log_e(k_{app})$ with r (Fig. 4) shows the drastic effect of [i-Bu$_3$Al] on the kinetics of propagation for values of r > 0.80. For r = 1, the system is unable to polymerize, suggesting the quantitative formation of a totally inactive 1/1 complex.

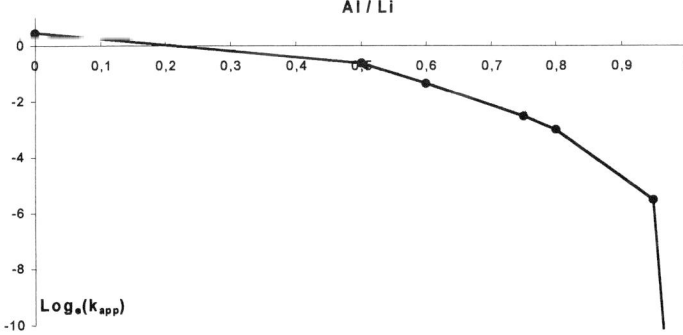

Fig. 4: Variation of $\log_e(k_{app})$ *vs* [Al]/[Li]; T = 50°C; cyclohexane.

MALDI TOF analysis of polystyrene samples obtained from i-Bu$_3$Al and PSLi seeds initiated by n-hexyllithium shows only the presence of chains bearing hexyl moieties (Fig. 5). This

156

supports that monomer insertion proceeds only in R-Li bonds to the contrary of n,s-dibutyl magnesium-based binary systems[3].

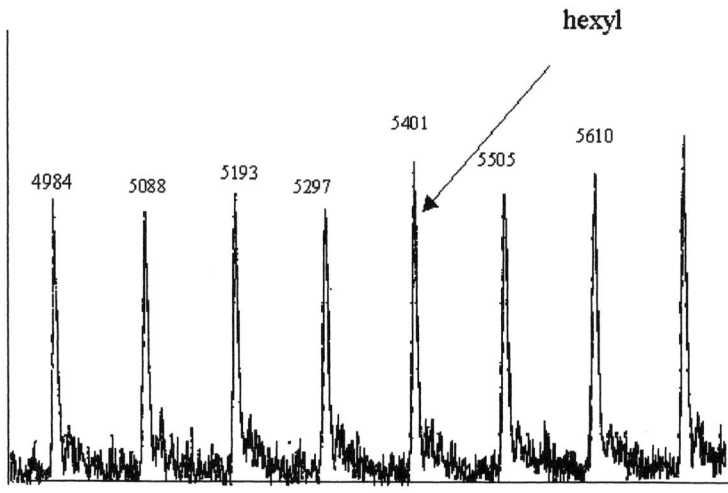

Fig.5: MALDI-TOF spectrogram corresponding to chains initiated by a hexyl moiety.

c) Persistence of Propagating Species

It is generally considered[10,11] that the thermal degradation of polystyryllithium in hydrocarbon solvents proceeds *via* the mechanism established by Szwarc and coll.[12] for polystyrylsodium in THF:

The rate of the degradation process can be determined from the disappearance of the initial U.V. absorption peak in the range 320-340nm (step 1) and the appearance of the new absorption peak in the range 450-500 nm (step 2). The rate of degradation is highly dependent on the dielectric constant of the medium. For example, the reaction is almost instantaneous for PS⁻, Na⁺ in hexamethylphosphortriamide (HMPA) at room temperature[12] whereas the half-life time of PSLi in cyclohexane at 100°C is around 3 hours.

Addition of R_2 Mg or R'_3Al to the living system increases strongly the persistence of the species (Fig. 6). Moreover, the absence of any U.V. signal in the 450-500nm range indicates that if the β-elimination of a hydride is possible (1ˢᵗ step), "ate" species are apparently unable to attack the acidic phenylallyl hydrogen generated in this step. This corroborates the lack of detectable "free" PSLi in equilibrium with the "ate" complex and the much lower reactivity of the "ate" complex propagating species than that of "free" PSLi.

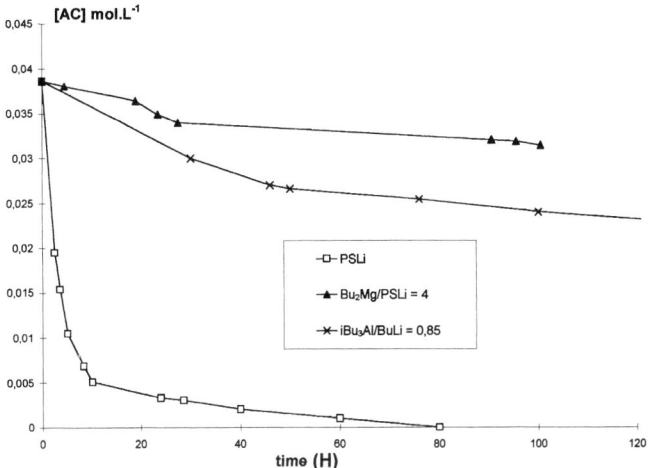

Fig 6: Variation of the concentration in active centers (AC) against time for different PSLi at 100°C and in cyclohexane.

d) Viscometric Studies

In order to obtain more information about the active "ate" complexes, the evaluation of the average number of polystyryl chains forming these complexes was obtained from viscometric measurements on dilute solutions. The degree of aggregation n can be derived from the ratio of the viscosity of active species to that of deactivated ones. It is given by the relation

$$n = (\eta_{act}/\eta_{deact})^2$$

η_{act} and η_{deact} being the relative viscosity of active and deactivated cyclohexane solutions, respectively.

From Table 2 it can be seen that η_{exp} varies from 1.80 to 1.05 for r varying from 0 to 1. This might indicate that the main part of the new species generated by addition of i-Bu$_3$Al form 1:1 complex. The formation of a 2:1 complex in small proportion (with 2 polystyryl chains) and of lower stability, in equilibrium with the 1:1 complex remains however possible.

Table 2: Determination of the aggregation state from viscosity measurements. $M_{n\,PSLi} = 5000g/mol$; $T = 35°C$; solvent: cyclohexane.

System	$n = (\eta_{act}/\eta_{deact})^2$
PSLi alone	1.80
PSLi + i-Bu$_3$Al (r = 0.25)	1.65
PSLi + i-Bu$_3$Al (r = 0.50)	1.43
PSLi + i-Bu$_3$Al (r = 0.75)	1.26
PSLi + i-Bu$_3$Al (r = 0.90)	1.11
PSLi + i-Bu$_3$Al (r = 1.00)	1.05

e) Nature of Propagating Species and Equilibria Involved

For r=1, all data point to the formation of the well known lithium tetraalkylaluminate complex, Li[Al(iBu)$_3$PS]. For values of r < 1 the kinetic behavior cannot be explained only by the progressive formation of the inactive 1/1 complex. The existence of equilibria between complexes with different stoichiometries (each having their own reactivity) in rapid exchange, is suggested. For example :

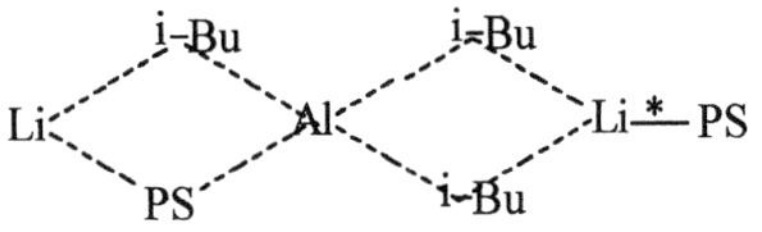

(corresponding to compounds b or b')

In this complex, the PS*-Li bond would be the only active part, as suggested by Arest-Yakubovitch for an Al/Na complexes[9]. The formation of the 2:1 complex of reduced reactivity would fit to the strong reactivity reduction at r > 0.5. In this range, the equilibrium

$$Li[Al(iBu)_3PS'] + Li_2[Al(iBu)_3(PS)_2] \rightleftharpoons Li_2[Al(iBu)_3(PS)(PS')] + Li[Al(iBu)_3PS]$$

would support the living character and the kinetic behavior of the polymerization. The aging process might involve the formation of the 2:1 complex from the mixture 1:1 complex and PSLi. Thus, we suggest the following scheme :

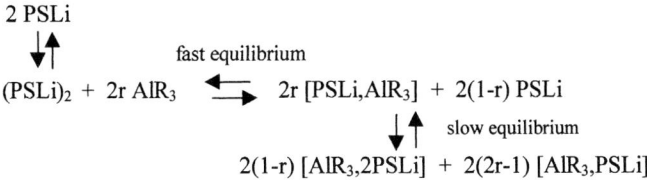

Experimental

Materials, techniques of polymerization and of characterization have already been described in the previous papers[1-3]. Additionally, triisobutylaluminium (ether-free, 1.0M in hexanes from Aldrich) was used as received. MALDI-TOF measurements were performed on a BIFLEX III (Bruker-Franzen Analytik GmbH) in reflection mode using a dithranol matrix in THF and silver trifluoroacetate. Viscosimetric measurements were performed in a Ubbelhode viscosimeter. The apparatus was placed inside a thermostated oven and a capillary (l = 10cm, d = 1mm) attached to the glass reactor was used.

Conclusion

The use of trialkylaluminium derivatives as an additive in the styrene anionic polymerization initiated by lithium derivatives allows a strong reduction of the reactivity of propagating active species in hydrocarbon media. The rate decrease depends upon the ratio [Al]/[Li] and a concomitant stability increase of the active species is observed. The living character of the polymerization, especially the control of molar masses is preserved over the entire range [Al]/[Li] studied. The elementary mechanisms involved in {triisobutylaluminium/polystyryllithium} styrene polymerization implies the formation of different mixed aggregates in rapid interchange, the final 1:1 one being inactive. Evolution of the [Al]/[Li] ratio varies the proportion of these species and allows to modulate the reactivity from a common lithiated polymerization to a completely dormant system. The exact nature of these species cannot yet be elucidated. We suggest a rapid equilibrium among inactive 1:1 and reactive 1:2 species. More detailed kinetic study is in progress.

The possibility to use these systems in styrene bulk polymerization will be presented in following papers.

References

1. P. Desbois,M. Fontanille, A. Deffieux, V. Warzelhan, S. Lätsch, C. Schade, *Macromol. Chem. Phys.*, **200**, 621 (1999)

2. P. Desbois, M. Fontanille, A. Deffieux, V.Warzelhan, S. Lätsch, C. Schade, *Ionic Polymerizations and RelatedProcesses*, Kluwer Academic Publishers, 223-237 (1999)

3. P. Desbois, M. Fontanille, A. Deffieux, V. Warzelhan, S. Lätsch, C. Schade, to be published

4. F. J. Welch, *J. Am. Chem. Soc.*, **81**, 1345 (1959)

5. C.B. Tsvetanov, D.T. Petrova, P.H. Li, M. Panayotov, *Eur.Pol.J.*, **14**, 25 (1978)

6. D.G.H. Ballard, R.J. Bowles, D.M. Haddelton, S.N. Richards, R. Sellens, D.L. Twose, *Macromolecules*, **25**, 5907 (1992)

7. K. Ute, T. Asada; Y. Nabeshima, K. Hatada, *Polym.Bull.*, **30**, 171 (1993)

8. H. Schlaad, A.H.E. Müller, *Macromol.Symp.*, **95**, 13 (1995)

9. A.A. Arest-Yakubovich, *Chem. Reviews*, **19**(4), 1-72 (1994)

10. M.D. Glasse, *Prog. Pol. Sci.* **9**, 133 (1983)

11. F. Schué, P. Nicol, R. Aznar, *Macromol. Chem., Macrom. Symp.* **67**, 213, (1993)

12. M. Fontanille, unpublished results

ANIONIC SYNTHESIS OF WELL-DEFINED POLYMERS WITH AMINE END GROUPS

Roderic P. Quirk* and Youngjoon Lee

717 Goodyear Polymer Center, Maurice Morton Institute of Polymer Science

The University of Akron, Akron, Ohio 44325-3909 USA

SUMMARY: The results for efficient tertiary and primary amine functionalization of polymeric organolithium compounds in hydrocarbon solution at room temperature are described for termination reactions with N-trimethylsilylbenzophenone imine, 3-dimethylaminopropyl chloride and 2,2,5,5-tetramethyl-1-(3-chloropropyl)-1-aza-2,5-disilacyclopentane. Functionalizations with the functionalized initiator, 2,2,5,5-tetramethyl-1-(3-lithiopropyl)-1-aza-2,5-disilacyclopentane are presented. Conditions for quantitative amine functionalization were observed for all of these reactions and reagents.

Introduction

One of the unique and useful aspects of alkyllithium-initiated, living anionic polymerization is the ability to prepare well-defined, ω-functionalized polymers by post-polymerization reactions with electrophilic reagents [1-3]. A particular challenge has been to develop methods that work in hydrocarbon solution at room temperature or at higher temperatures, i.e. under conditions in which polydienes with high 1,4 microstructure are obtained [2]. Although a variety of functionalization reactions have been reported, many of these specific functionalization reactions are either inefficient or have not been adequately characterized. A particular challenge has been to prepare primary amine-functionalized polymers [1-9], while avoiding complex, multistep processes [10].

In general, it is necessary to protect the primary amine group because of the acidity of the amine protons; for example, the pK_a of cyclohexylamine has been estimated to be 41.6 [11]. Herein, the scope and limitations of living anionic functionalization methods for synthesis of amine-functionalized polymers are delineated with emphasis on new, efficient procedures. All polymerization and functionalization reactions have been carried out in all glass, sealed reactors using breakseals and standard high vacuum techniques [12].

Reactions of polymeric organolithium compounds with N-trimethylsilyl-protected benzophenone imine

$$
\text{PLi} + \underset{C_6H_5}{\overset{C_6H_5}{>}}\!\!=\!\!NSi(CH_3)_3 \longrightarrow \left[\begin{array}{c} C_6H_5 \\ | \\ Li \\ P\!-\!C\!-\!NSi(CH_3)_3 \\ | \\ C_6H_5 \end{array} \right] \quad (1)
$$

1

$$\downarrow H_3O^+$$

$$
\begin{array}{c} C_6H_5 \\ | \\ P\!-\!C\!-\!NH_2 \\ | \\ C_6H_5 \end{array} \quad \mathbf{2}
$$

One approach for the anionic synthesis of primary amine-functionalized polymers is based on the reaction of polymeric organolithium compounds with N-trimethylsilyl-protected imines [4,5,7]. In order to effect efficient functionalization, the protected imine should not contain acidic alpha hydrogens (enolization-like side reactions) [4,5], nor should they be derivatives of aldehydes (Cannizzarro-like side reactions) [7]. Following this rationale, the N-trimethylsilyimine derivative of benzophenone (1) was prepared [13] and investigated as a functionalization agent for preparation of primary amine functionalized polymers (2) by reaction with polymeric organolithium compounds as shown in Eq. (1). The reaction of poly(styryl)lithium with **1** was effected in benzene solution at room temperasture using both normal (addition of **1** to PSLi) and inverse addition procedures. The amine functionalization of PSLi by reaction with **1** required 8 hours in benzene solution using the inverse addition (PSLi added to **1**). In the presence of 5 vol % THF (added after the polymerization) the functionalization reaction was competed in 2 hours. For both of these functionalization reactions, no non-functional polymer was observed by TLC analysis, the functionalized polymer was isolated in > 99 % yield by silica gel column chromatography and the functionalities determined by end-group titration were 93 % and 97 % in benzene and in the presence of THF, respectively. It is noteworthy that no dimer formation was observed, since dimer formation was observed for functionalizations with N-(benzylidene)-trimethylsilylamine [7]. The relatively slow rate of addition of PSLi to **1** is presumably due to the steric congestion which develops in the transition state for the reaction. A Newman projection[14] of the product **2** illustrating the unfavorable gauche-gauche interactions is shown below.

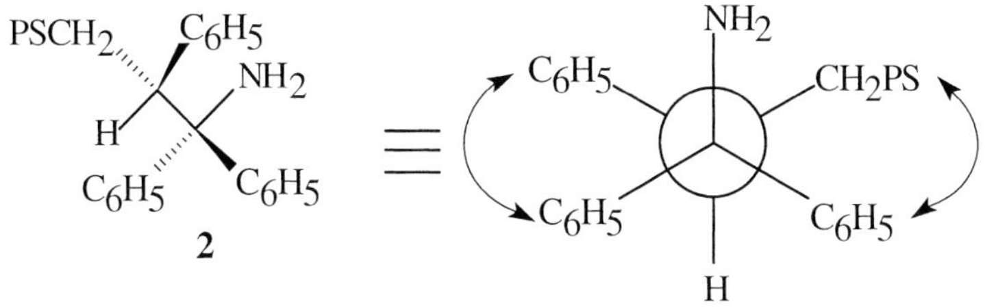

Reactions of polymeric organolithium compounds with amine-substituted alkyl chlorides

The reaction of polymeric organolithium compounds with substituted alkyl chlorides has been investigated as a general functionalization reaction as shown in Eq. (2) [15]. Although high functionalization yields have been reported

$$PLi + Cl\text{—}CH_2CH_2CH_2\text{—}Z \longrightarrow P\text{—}CH_2CH_2CH_2\ Z \qquad (2)$$

$$Z = \text{-OR, NR}_2\text{, -SR, etc.}$$

for such functionalization reactions at low temperatures in THF [8,16,17], a systematic study of this reaction showed that at room temperature and above in hydrocarbon solution, the reaction is complicated by lithium-halogen exchange leading to polymer-polymer coupling reactions (dimer formation) and by hydrogen abstraction from the β-hydrogen of the alkyl chloride (formation of unfunctionalized polymer) [15]. The results for functionalization of poly(styryl)lithium are shown in Eq. (3). In conjunction with a comprehensive

$$PSLi + Cl—CH_2CH_2 \; CH_2N(CH_3)_2 \xrightarrow[\text{25 °C}]{C_6H_6} PS—CH_2CH_2CH_2N(CH_3)_2 +$$

$$67 \; \%$$

$$PS—PS \; + \; PS—H \quad \quad (3)$$

$$23 \; \% \quad \quad \quad 10 \; \%$$

study of experimental variables which could increase the efficiency of this and other functionalization reactions of polymeric organolithiums with substituted alkyl chlorides, the effect of the addition of lithium halides has been investigated. Previous work has indicated that cross-associated aggregates of alkyllithium compounds and lithium halides [18] generally have decreased reactivity, and presumably higher selectivity compared with the uncomplexed alkyllithiums [19-22]. Therefore, it was of interest to determine if added lithium chloride would promote the functionalization reactions of polymeric organolithium compounds with 3-dimethylaminopropyl chloride in hydrocarbon solution. Quite unexpectedly, it was found that ω-dimethylamino-polystyrene was obtained in quantitative yield from the functionalization of poly(styryl)lithium with 3-dimethylaminopropyl chloride in the presence of one equivalent of lithium chloride. No dimer formation was observed by SEC and amine end-group titration was consistent with quantitative functionalization. No unfunctionalized polystyrene was detected by TLC analysis and the amine-functionalized polystyrene was isolated in quantitative yield by silica gel column chromatography.

In order to determine the generality of the effect of lithium chloride on functionalization reactions with substituted alkyl chlorides, the reaction of poly(styryl)lithium with 2,2,5,5-tetramethyl-1-(3-chloropropyl)-1-aza-2,5-disilacyclopentane (3) has been investigated at room temperature in hydrocarbon solution as shown in Eq. (4). In the absence of lithium chloride, SEC analysis showed the presence of dimer (10 %) and the amine functionality of the product was only 77 % as determined by end-group titration. However, in the presence of 1.5 equivalents of lithium chloride, no dimer formation was observed by SEC and the functionality was 99 % as determined by end-group titration. No non-functionalized polymer was detected by TLC analysis. With respect to the sensitivity of the TLC method for detection of non-funcitonalized polymer, previous studies have shown that this technique is capable of detecting 1-2 wt. % levels of non-functionalized polymers [9]. These results indicate

$$PSLi + Cl{-}CH_2CH_2CH_2{-}N\left[\begin{array}{c} \overset{CH_3}{\underset{CH_3}{Si}}\\ \overset{CH_3}{\underset{CH_3}{Si}} \end{array}\right]_{\!3} \xrightarrow[C_6H_6]{25\,°C\ \ CH_3OH} PS{-}CH_2CH_2CH_2NH_2 \qquad (4)$$

that lithium chloride is effective in promoting the coupling of polymeric organolithiums with substituted alkyl chlorides. Since these reactions are efficient in hydrocarbon solution at room temperature and above and since a variety of substituted alkyl chlorides are readily available, this methodology for chain-end functionalization has the potential to be the most general and useful method for chain-end functionalization of polymeric organolithium compounds.

Functionalization with functionalized initiators.

A simpler, quantitative functionalization methodology utilizes functionalized alkyllithium initiators [1,23]. There are several distinct advantages in the use of a functionalized initiator. For alkyllithium-initiated polymerization, each functionalized initiator molecule will produce one macromolecule with a functional group from the initiator residue at the initiating (α) chain end and with the active carbanionic propagating species at the terminal (ω) chain end regardless of molecular weight as shown in Eq. (5).

$$\textbf{prot-N-}RLi + n\ monomer \xrightarrow[\geq 25\,°C]{RH} \textbf{prot-N-}PLi \qquad (5)$$

Because most functional groups of interest (e.g., hydroxyl, carboxyl, amino) are not stable in the presence of either simple or polymeric organolithium reagents, it is generally necessary to use suitable protecting groups in the initiator [1]. A suitable protecting group is one that is not only stable to the anionic chain ends but is easily removed upon completion of the polymerization to generate the desired functional group.

The utility of alkyllithium initiators with protected hydroxyl groups (*t*-alkoxy and *t*-butyldimethylsiloxy) has been investigated for preparation of functionalized polymers (monofunctional, telechelic, heterotelechelic and functionalized, star-branched) in hydrocarbon solution [24,25]. Like hydroxyl groups, primary and secondary amine groups are not stable in the presence of organolithium chain ends; they undergo proton transfer reactions with the active chain ends [26]. As discussed previously, a suitable protecting group for a primary amine is the corresponding *bis*(trialkylsilyl) derivative [8,16]. Schulz and Halasa [27] prepared *p*-lithio-N,N-*bis*(trimethylsilyl)aniline (**4**) and investigated its use as a protected amine initiator for the anionic polymerization of butadiene and isoprene. Unfortunately, like many other functionalized alkyllithium initiators, **4** is not soluble in hydrocarbon solvents. Therefore,

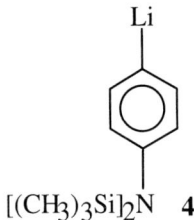

$[(CH_3)_3Si]_2N$ **4**

it was necessary to prepare the initiator in diethyl ether and to effect polymerizations in mixtures of hexane and diethyl ether. Polybutadienes with relatively narrow molecular weight distributions and functionalities of 0.69-1.0 were obtained using this initiator; however, the vinyl microstructures of these polydienes ranged from 39-50 %. A hydrocarbon-soluble amine-functionalized initiator is required to prepare amine-functionalized polydienes with high 1,4-microstructure [2].

A potentially useful protected amine-functionalized initiator is 2,2,5,5-tetramethyl-1-(3-lithiopropyl)-1-aza-2,5-disilacyclopentane (**5**). This initiator is soluble in cyclohexane, a necessary but not sufficient condition for a useful alkyllithium initiator [24]. However, unlike previous studies with protected hydroxyl-functionalized initiators such as 3-(*t*-butyldimethylsiloxy)propyl-lithium (**6**) and 3-(*t*-butoxy)propyllithium (**7**) [24,25], this initiator did not initiate polymerization of isoprene in either cyclohexane or benzene, at room

Structures 5, 6, 7:

$$\underset{5}{\text{(Me}_2\text{Si)}_2\text{NCH}_2\text{CH}_2\text{CH}_2\text{Li}} \quad \underset{6}{(\text{CH}_3)_3\text{SiOCH}_2\text{CH}_2\text{CH}_2\text{Li}} \quad \underset{7}{(\text{CH}_3)_3\text{COCH}_2\text{CH}_2\text{CH}_2\text{Li}}$$

temperature or at 60 °C. These results suggest that there are some specific interactions of the lithium centers in the corresponding aggregated species which stabilize the initiator, making it less reactive. The x-ray crystal structures of both 3-lithio-1-methoxybutane (**8**) [28] and 3-dimethylamino-1-lithiopropane (**9**) [29] show tetrameric structures for both of these organolithiums with intramolecular coordination of the heteroatom to lithium as indicated schematically below. It is interesting to note that 3-dimethylamino-1-lithiopropane is also tetrameric in benzene solution [30], whereas most primary alkyllithiums such as *n*-butyllithium and ethyllithium are hexameric in hydrocarbon solution [2]. These results suggest that intramolecular stabilizing interactions can occur when there are heteroatoms available for coordination as substituents on the 3-carbon of the organolithium. Thus, such interactions may be responsible for the reduced reactivity of **5** as an initiator for isoprene polymerization.

Another factor that may contribute to the reduced reactivity of initiator **5** is the ability of silicon to inteeract with nucleophiles by expanding it octet to form a pentacoordinated siliconate intermediate [31]. To the extent that coordination of the carbanionic chain end to a silicon atom in the protecting group, the chain end could be stabilized and thus be less reactive.

Fortunately, it was possible to active this initiator towards isoprene polymerization by addition of one equivalent of triethylamine at 60 °C as shown in Eq. (6).

$$(6)$$

Under these conditions, the observed molecular weight (M_n = 8,300 g/mol) was higher than the calculated value (M_n = 7,000 g/mol) and the molecular weight distribution was narrow (M_w/M_n = 1.08). The observed molecular weight corresponds to an initiator efficiency of only 84 %. The polyisoprene microstructure was 85 % 1,4. Only one spot was observed by TLC analysis of the product. Furthermore, and most importantly, the amine functionality was determined to be 1.06 by amine end group titration. It is noteworthy that no polymerization was observed at room temperature in the presence of only one equivalent of triethylamine. Further work is in progress to evaluate the usefulness and applications of this protected amine-functionalized initiator.

References

1. *Functional Polymers: Modern Synthetic Methods and Novel Structures*, A. O Patil, D. N. Schulz, B. M. Novak (Eds.), *ACS Symposium Series*, **704** (1998)
2. H. L. Hsieh, R. P. Quirk, *Anionic Polymerization: Principles and Practical Applications*, Marcel Dekker, New York, 1996
3. R. P. Quirk in: *Comprehensive Polymer Science*, First Supplement, S. L. Aggarwal and S. Russo (Eds.), Pergamon Press, Elmsford, New York, 1992, p. 83
4. A. Hirao, I. Hattori, T. Sasagawa, K. Yamaguchi, S. Nakahama, *Makromol. Chem., Rapid Commun.* **3**, 59(1982)
5. 5. I. Hattori, A. Hirao, K. Yamaguchi, S. Nakahama, N. Yamazaki, *Makromol. Chem.*, **184**, 1355(1983)
6. R. P. Quirk and P.-L. Cheng, *Macromolecules*, **19**, 1291(1986)
7. R. P. Quirk and G. J. Summers, *Brit. Polym. J.*, **22**, 249(1990)
8. K. Ueda, A. Hirao, S. Nakahama, *Macromolecules*, **23**, 939(1990)
9. R. P. Quirk and T. Lynch, *Macromolecules*, **26**, 1206(1993)

10. J. J. Cernohous, C. W. Macosko and T. R. Hoye, *Macromolecules*, **31**, 3759(1998)

11. A. Streitwieser, Jr., E. Juaristi, L .L. Nebenzahl in *Comprehensive Carbanion Chemistry*, Part A, E. Buncel and T. Durst, Eds., Elsevier, Chapt. 7, 1980, p. 323.

12. M. Morton and L. J. Fetters, *Rubber Chem. Technol.*, **48**, 359(1975)

13. C. Kruger, E. G. Rochow, U. Wannagat, *Chem. Ber.*, **96**, 2132(1963)

14. E. L. Eliel and S. H. Wilen, *Stereochemistry of Organic Compounds*, Wiley-Interscience, New York, 1994, p. 1202.

15.   R. P. Quirk, K. Han and Y. Lee, *Polym. Internat.*, **48**, 99(1999)

16. K. Iwasaki, A. Hirao, S. Nakahama, *Macromolecules*, **26**, 2126(1993)

17. P. Charlier, R. Jerome, P. Teyssie, *Macromolecules*, **25**, 617(1992)

18. D. P. Novak, T. L. Brown, *J. Am. Chem. Soc.*, **94**, 3793(1972)

19. W. Tochtermann, *Angew. Chem. Int. Ed. Eng.*, **5**, 351(1966)

20. W. Glaze and R. West, *J. Am. Chem. Soc.*, **83**, 4437(1960)

21. R. Waack, M. A. Doran, *Chem. and Ind. (London)*, 496(1964)

22. R. Fayt, R. Forte, C. Jacobs, R. Jerome, T. Ouhadi, P. Teyssie, S. K. Varshney, *Macromolecules*, **20**, 1442(1987)

23. R. P. Quirk, S. H. Jang, J. Kim, *Rubber Chem. Technol.* **69**, 444(1996)

24. R. P. Quirk, S. H. Jang, K. Han, H. Yang, B. Rix, Y. Lee in *Functional Polymers: Modern Synthetic Methods and Novel Structures*, A. O Patil, D. N. Schulz, B. M. Novak (Eds.), *ACS Symposium Series*, **704** (1998), p. 71

25. R. P. Quirk, S. H. Jang, H. Yang and Y. Lee, *Macromol. Symp.* **132**, 281(1998)

26. W. C. E. Higginson and N. S. Wooding, *J. Chem. Soc.*, 769(1952)

27. D. N. Schulz and A. F. Halasa, *J. Polym. Sci., Polym. Chem. Ed.*, **14**, 2401(1977)

28. G. W. Klumpp, P. J. A. Geurink, A. L. Spek, J. M. Duisenberg, *J. Chem. Soc. Chem. Commun*, 814(1983)

29. K. S. Lee, P. G. Williard, J. W. Suggs, *J. Organomet. Chem.*, **299**, 311(1986)

30. K.-H. Thiele, E. Langguth, G. E. Muller, *Z. Anorg. Allg. Chem.*, **462**, 152(1980)

31. R. Corriu in *Silicon Chemistry*, E. R. Corey, J. Y. Correy, P. P. Gaspar, Eds., Wiley, 1988, p. 225.

Macromol. Symp. 157, 171–182 (2000) 171

Synthesis of PMMA and PMMA Block Copolymers at Elevated Temperatures by Phosphor Ylide-Mediated Polymerizations

Dimo K. Dimov , William N. Warner, and Thieo E. Hogen-Esch*

Loker Hydrocarbon Research Institute, Department of Chemistry
University of Southern California, Los Angeles, CA 90089, U.S.A.

Stephan Juengling and Volker Warzelhan

Polymer Research Laboratory, ZKT-B1, BASF AG, 67056
Ludwigshafen, Germany

SUMMARY: The MMA polymerization initiated by (1-naphthyl) triphenylphosphonium triphenylmethyl anion(TPM,NTPP) or the NTPP salt of the methylisobutyrate anion (MIB,NTPP) in THF at temperatures varying from 25 to 70^0C appears to be a living polymerization as indicatred by a linear M_n vs. MMA conversion plot and by the narrow MW distributions. As indicated by the proton spectrum of MIB, NTPP in THF d_8 the predominant polymerization intermediate appears to be the ylide formed exclusively by addition of the MIB anion to the 4-carbon of the 1-naphthyl group. This ylide shows upfield shifts for the 3- proton of the cyclohexadienyl ring and of the 6- and 8--protons of the remaining aromatic ring.

The rates of the NTPP mediated polymerizations are reduced by factors of around 10^4-10^5 compared to that mediated by the corresponding ylides formed by addition of MIB to the phenyl ring of tetraphenylphosphonium ion. The reductions in rate in the presence of NTPP make it possible to carry out MMA polymerizations under conditions not normally accessible.

Introduction

There has been a great deal of interest in the controlled polymerization of (meth)acrylates at ambient conditions in recent years.[1,2] We have recently reported tetraphenylphosphonium carbanion initiators that polymerize MMA at temperatures between 0 and 25^0 C to give polymers of high molecular weights (MW • 50,000) and narrow polydispersities.[3,4] However these polymerizations were shown to be very rapid with monomer conversion half life times in the order of one second at 25^0C under typical reactions conditions and have the disadvantage that thermal control is limited due to the exothermicity of the polymerization.

 CCC 1022-1360/00/$ 17.50+.50/0

We now report the 1-naphthyltriphenylphosphonium triphenylmethyl anion (NTPP, TPM) initiated living anionic polymerization of MMA in THF at temperatures from 25 to 70^0C. These polymerizations are much slower then that in the presence of the tetraphenylphosphonium (TPP) ion and at 25^0C have half lives in the order of several hours. Because of the much lower rates these polymerizations can be carried out at or below 70^0C but all indications are that the polymerization temperatures can be increased further. Molecular weight distributions are narrow even at 70^0C with polydispersities of less than 1.20.

Experimental

Triphenylmethylpotassium (TPMK) was prepared from triphenylmethane and potassium metal in THF or from the reaction of the potassium dianion oligomer of alphamethylstyrene with triphenylmethane.   Potassiomethylisobutyrate (KMIB) was synthesized by metalation of methylisobutyrate with hexamethyldisilazylpotassium.   1-Naphthyltriphenyl-phosphonium bromide (NTPPBr) was prepared by stirring $NiBr_2$ · $2Ph_3P$ and 1-bromonaphthalene overnight at 190°C under argon.[5] The removal of traces of water from the salt was carried out by titration with TPMK.[6] Thus a solution of Ph_3CK was added to the NTPPBr salt until a reddish color of the triphenylmethyl salt was clearly visible. The salt was then washed with THF until the red color had disappeared and was then dried overnight under high vacuum.

Initiators were prepared by addition of TPMK to a slurry of the NTTPBr in THF at or below room temperature using procedures similar to that reported previously.[3,4] A rapid formation of the NTPP carbanion could be observed by a slight change in color from bright red to a deep tomato red. The metathesis reaction could be observed by the precipitation of KBr that was not filtered off as it did not interfere with the polymerizations. The system was then warmed to the desired temperature and stirred for 15 minutes to allow complete cation exchange. The longer metathesis time was found to be necessary since the exchange reaction was slower than for the case of the tetraphenylphosphonium salt (TPPBr). The slower metathesis reaction is at least partially attributable to the statistical factor of four that favors the ylide formation in the TPP system.

Polymerizations were carried out by a one time addition of a THF solution of MMA through a teflon stopcock at or below room temperature followed by warming to the desired reaction temperature and the polymerization mixture was stirred for the specified reaction time (Table 1).

Alternatively the polymerization was carried out under inert gas. In one case (Table 1#12) the polymerization of MMA was carried out by a one time addition of MMA (10 mL) to the TPM, NTTP solution dissolved in THF to give a concentrated MMA solution (50 weight %). In this case the exothermic polymerization caused a rise in temperature to 76^0C causing the THF to reflux. The reaction in this case was complete in about 30 minutes and ,after addition of 50 mL of THF, was terminated by addition of 1 mL of methanol/acetic acid(1/1 V/V).

A PMMA -poly(t-butylmethacylate) block copolymer was synthesized by sequential addition of MMA (0.89M) and TBMA (0.62M)to the TPM, NTTP initiator at ambient conditions. In this case the reaction temperatures varied between 25 and 50^0C.

Polymer yields were determined gravimetrically in some cases by precipitation in hexane followed by drying.Polymer isolation, SEC, and NMR procedures have been described elsewhere.[3,4]

Results and Discussion

As was observed in the case of the TPP systems the initiation of MMA in the presence of the NTPP cation is indicated by a rapid change from deep red color to an orange brown the color of which is slightly different from that of the corresponding TPP ylide. The polymerizations of MMA in THF in the presence of NTPP counterion proceed in a controlled manner at temperatures from $20°$ up to $76°$C. No upper limit for the polymerization was explored. The polymerization rates are *much* slower -on the order of 30 minutes to 24 hours- than observed for the TPP-mediated polymerizations that are typically complete in about one second. Clearly polymerization rates are four to five orders of magnitude lower than that determined in the TPP systems.[7]

The data shown in Table 1 are consistent with a living polymerization even at 70°C and good molecular weight control as indicated by a linear increase in M_n with conversion using a method by Penczek.[8] (Figure 1). Also relatively narrow MW distributions were obtained even at high temperatures (Table 1).

The termination reactions of the NTPP-mediated systems by water or alcohols typically can take up to several minutes or longer and are much slower than the TPP systems. Also upon opening the system to air, decoloration takes far longer than for the TPP systems.These observations are clearly related to the greater stability of the corresponding naphthyl ylide the stability of which is enhanced by the unperturbed aromaticity of the remaining benzene moiety (see below).

Table 1. NTPP$^+$,Ph$_3$ C$^-$ - Mediated Polymerization of MMA in THF at Various Temperatures.[f]

Run	Rxn Time[a]	Rxn Temp (°C)	Yield (%)	M$_n$ (SEC)	PDI (SEC)	f[b]
1	28 min	25	-	1,250	1.05	-
2	28/19 min	25-70	-	2,500	1.03	-
3	28/49 min	25-70	-	3,800	1.11	-
4	70 min	70	40	5,700	1.06	0.26
5	3 hrs	70	50	9,300	1.17	0.25
6	20 hrs	70	> 95	19,000	1.17	0.25
7	3 hrs	25	80	8,300	1.08	0.40
8	12 hrs	70	> 95	9,900	1.07	0.40
9	4 hrs	70	> 95	15,000	1.12	0.24
10	20 hrs	70	> 95	17,000	1.14	0.23
11	18 hrs	25	> 95	7,300	1.24	0.36
12[d]	30 min	25-76[c]	> 95	24,100	1.23	0.66
13[e]	30 min	25-50	>95	9,500	1.14	-

a.Dual times indicate at 25 and 70°C respectively b) Initiator efficiency, M$_n$ (Calc) / M$_n$ by SEC using PMMA standards. c. Reaction temperature increase by adiabatic heating from 25 to 76°C. d. Syndiotactic content is 53%. e. Block copolymerization of MMA (0.89M) at 25-35°C and t-BuMA (0.62M) at 50°C for 24 hrs. f. MMA concentration is 0.16M

The initiator efficiencies for the NTPP,TPM system in THF ranged from 20% to 66%, regardless of reaction temperature. It is worth pointing out that these are calculated from the concentrations of the TPMK and similar precursor anions and not from the actual concentrations of the NTPP carbanion. Thus it is likely that these low values are apparent only and due to a combination of factors including traces of water present in the very hygroscopic phosphonium salts, to some TPMK decomposition and similar inadvertent loss of anion concentration prior to MMA addition.

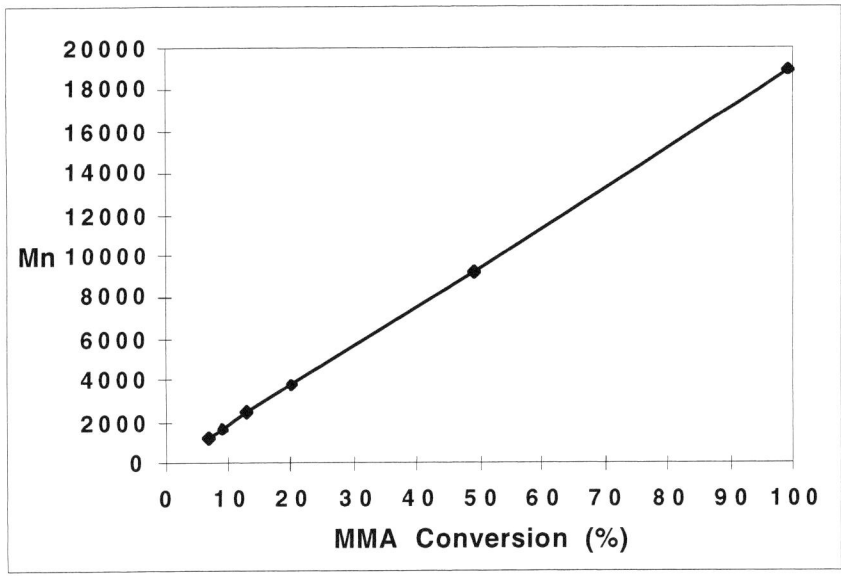

Figure 1. TPM,NTPP initiated polymerization of MMA. Number-average MW as a function of conversion.

The stereochemistry of the polymerizations is similar to that of anionic or radical polymerizations carried out at similar elevated temperatures.[9] Thus a predominantly syndiotactic stereochemistry was found but this does not provide a tool to elucidate the polymerization mechanism although in some cases this method was used to distinguish an ylide mediated polymerization from that in the presence of potassium ion.[6]

The occurrence of a living polymerization at very high monomer concentration (about 4.5M) and at relatively high temperatures is of interest (Table 1). For instance under these conditions for the case of run 12 the polymerization took only 30 minutes for completion and the heat of reaction gave a temperatures that caused the THF to start refluxing. Nevertheless the polymer had a relatively narrow distribution (Table 1). This would seem to indicate potential for industrial applications. The suitability of this NTPP mediated polymerization for the synthesis of PMMA-b-PTBMA block copolymers was demonstrated (Table 1,#12).

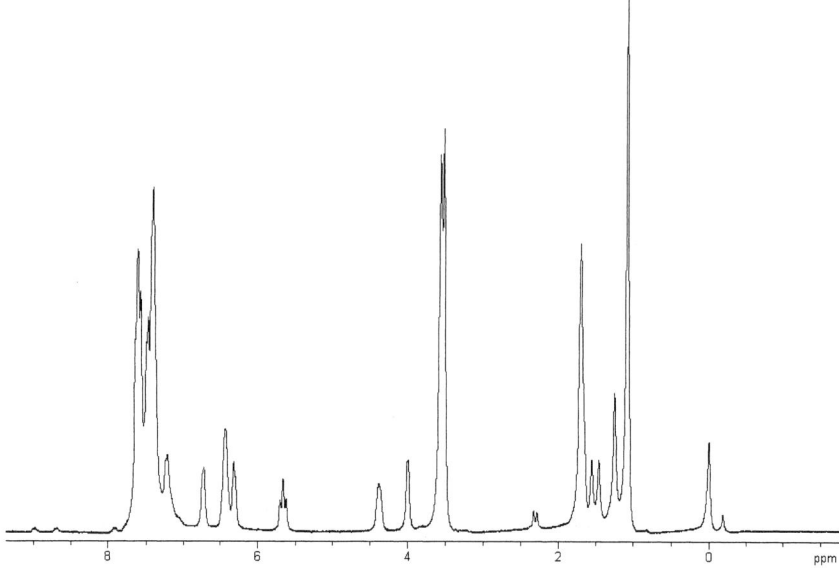

Figure 2. ^{1}H NMR (250 Mhz) spectrum of of MIB, NTPP (ylide 3) in d_8-THF).

Surprisingly the polymerization of acrylates with the NTTP system were unsuccessful. In fact the MW the distributions and the control of the polymerization was less favorable than obtained with the TPP system where at least there was some indication for potential in the synthesis of PMMA- polybutylacrylate block copolymers.[6]

Characterization. The presence of ylides in the very similar TPP cation mediated MMA polymerization was demonstrated by NMR studies of the TPP MIB derivative that was synthesized by the reaction of the potassium MIB anion with TPP chloride in THF at low temperature and these studies indicated that essentially all of the MIB is in the ylide form. [3,4,6] This ylide is generated by the addition of the MIB anion to one of the four phenyl rings of the TPP cation (Scheme 1).[10]

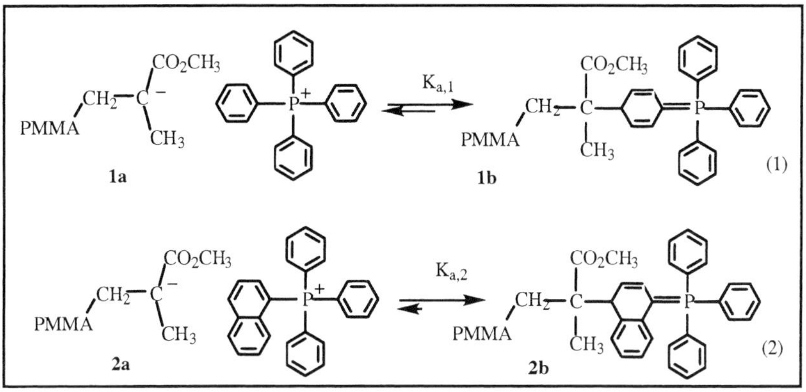

Scheme 1.   Ylide formation by addition of PMMA anion to TPP and NTPP cations.

The ^{1}H NMR of MIB,NTPP (**3**) that was synthesized by reaction of KMIB with NTPPBr in THFd8 displays a spectrum consistent with ylide formation by addition of the MIB anion exclusively to the 1-naphthyl ring of NTPP (Figure 2). As shown in the case of the TPP ylide the upfield shift of the cyclohexadienyl protons is consistent with the presence of the ylide. Thus proton (b) is present at about 4.4 ppm. The upfield shift of this proton is consistent with the expected negative charge on this carbon (Scheme-2). The upfield shifts of protons (e) and (g) are again consistent with the delocalization of negative charge into the aromatic ring as shown in Scheme -2. The upfield shifts of the remaining protons (a), d) and (f) are smaller consistent with Scheme 2.

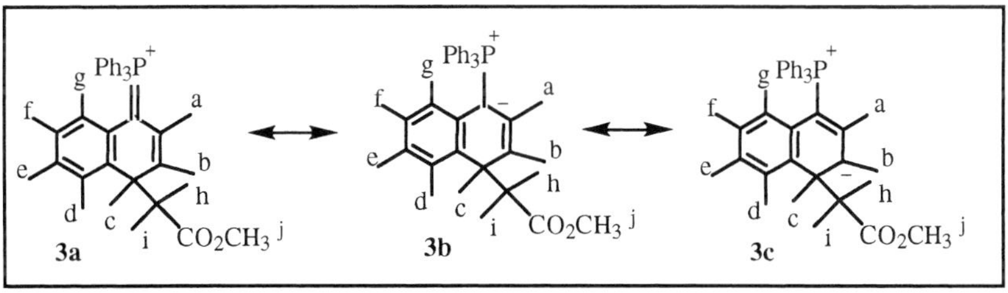

Scheme 2.   Resonance structures of ylide, **3**.

Methine proton (c) is observed at about 4.0 ppm. The downfield resonance of this benzylic proton attributable boht to the diamagnetic shielding of the aromatic ring and to the deshielding and electron withdrawing effect of the ester group. Although the methyl groups would be expected to be diastereotopic only a single resonance for protons (i) and (j) is observed. The methoxy protons show a resonance at about 3.6 ppm. The other resonances at 1.3, 1.5 and 2.3 ppm appear to be due to solvent impurities or other sources unrelated to the ylide and but are not consistent with the presence of NTPPBr or of related species such as **2a.**

Although there is no NMR evidence thus far for the existence of the ion pairs **2a** or **1a**, their presence should not be ruled out. First the detection of small quantities of ionic species is difficult to showby NMR. Second, if the interconversion of the ion pairs and ylides is fast on the NMR time scale the direct observation of the ion pairs may only be possible at very low temperatures. This possibility can not be ruled out and is even likely that ions or ion pairs are involved inthe polymerization (see below).

Mechanism of the polymerization. The above and previous results on the ylide mediated polymerizations of MMA appear to be first documented cases of vinyl polymerizations mediated by ylides.[3,4,6] In these cases ylides appear to be the predominant intermediates. A previous report has documented the involvement of an ylide in the initiation step in an otherwise anionic polymerization.[11] It should also be pointed out that in the presence of small amounts of water the mechanism of the TPM,TPP initiated polymerization readily reverts to what was shown to be an anionic mechanism.[6]

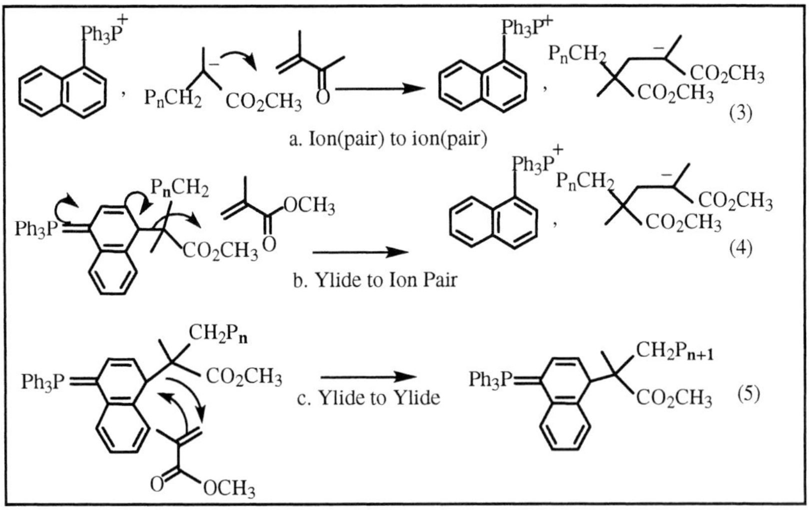

Scheme 3.   Possible MMA addition steps involving ylides and ion pairs.

The polymerization mechanism is of considerable interest particularly as it relates to that postulated for analogous polymerizations that are anionic or anion related.[1,2] The first question is the nature of the propagating species. It is clear that the predominant species in both the TPP and the NTPP systems is the ylide. In fact no other species could be shown conclusively by NMR to be present in addition to the NTTP ylide. As indicated above, this is not too surprising given the experimental limitations.

A key question in the mechanism of the polymerization is the nature of the propagating chain. As shown in Scheme-3 the propagation could be carried by ionic species (eqn.3) or by ylides (eqn.5). Another possibility is the addition of ylides to monomer giving ion pairs that could collapse back to ylides or react with monomer (eqn.4) (see below).

In the first (simplified) mechanism an unreactive ylide forms a NTPP ion pair in what could be called a tautomeric equilibrium (Scheme 1, eqn. (1)).[12] This ion pair would be expected to have a very high reactivity compared with the ylide first beacuse it is known that the major species is the ylide and second as the counterion in this case is known to be large. Thus the ion pair and the free ion would be expected to have similar reactivities.[9] If this situation prevails it would have to follow that the interconversion of ion pairs and ylides is fast compared to the polymerization time scale. This requirement is likely to be satisfied for the case of the NTPP ylides as polymerization half lives are on the order of hours. This situation may be different for the case of the TPP ylide where the polymerizations are much faster.

The enormous decrease (factor of 10^4- 10^5) in the value of the apparent polymerization rate constants compared to that of the corresponding TPP ylide-mediated polymerizations clearly reflects the much greater stability of the NTPP ylide. Thus the stability of ylide **2b** like that of its model, ylide **3**, is most likely due to the restoration of full benzene aromaticity in the naphthyl ylides upon addition of the enolate nucleophile to the naphthalene ring (Scheme -1). The TPP ylide **1b**, is expected to be less stable due to the disruption of the benzene aromaticity upon addition of the PMMA anion to the ring. Assuming an equilibrium between the ion pairs (**1a,2a**) and ylides (**1b, 2b**), the association constant $K_{a,2}$ is expected to be greater than $K_{a,1}$.

The less than full benzene aromaticity of the naphthyl rings is consistent with lower heats of hydrogenation and with the greater susceptibility of naphthalene toward nucleophilic attack.[13] As the resonance energies per ring in benzene and naphthalene are 36 and 61 Kcals, the restoration of full aromaticity to one of the two naphthalene rings is estimated as about 6 Kcal.[12] This is expected to add considerable driving force in the formation of the NTPP as opposed the TPP ylide. If the change in entropy is assumed to be the same for reactions (1) and (2) the decrease in free energy of of **2b** of 6 Kcals would increase the value of $K_{a,2}$ relative to $K_{a,1}$ by a factor of about 2.10^4. Given equilibria (1) and (2) the concentrations of ionic species would be expected to increase linearly with the inverse of the association constants so that the concentration of **1a** would be expected to be about a factor of about 10^4 larger than that of **2a.** It is interesting that this increase is approximately equal to the increase in reactivity of the TPP relative to the NTPP system and this would seem to support the view that the propagation is carried by dissociated ionic species. The ion pairs because of their very low concentrations under the polymerization conditions are to be dissociated into free ions to a considerable extent. This free ion would be expected to have a reactivity close to, but somewhat greater than that of the ion pair.

The above does not necessarily exclude ylides as the reactive intermediates. However reaction (5) would not appear to be likely. Thus no precedent for such a reaction has been shown. Although an ylide to ylide type polymerization (eqn. 5) is possible this seems unlikely as high level *ab initio* calculations have shown this reaction to be too high in activation energy.[14]

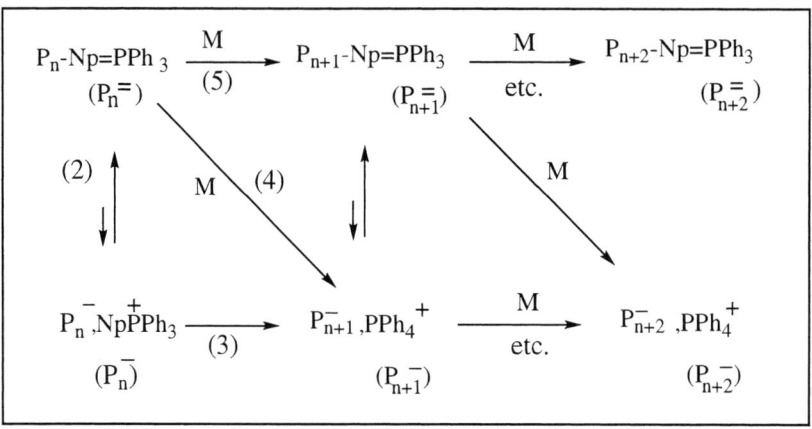

Scheme 4.   Possible polymerization mechanisms involving ylides as intermediates.

A more complicated situation would prevail for the cross-over reaction (4) that would lead to ylides and ion pairs that may both be contributing to the polymerization. This possibility should not be dismissed too readily but the corresponding polymerization is expected to be far more complicated. Because of its greater compexity experimental testing of this model would not be simple.

The corresponding mechanisms are summarized in Scheme -4. At this point it seems most likely that ionic species are the dominant if not the sole propagating species. First the ions (ion pairs) are higher in energy than the ylides for both of the ylide systems studied so far and are thus expected to be much more reactive than the ion pairs. Experiments to probe the mechanisms are underway.

Conclusions

The anionic polymerization of MMA in the presence of NTPP cation displays living character and is capable of producing PMMA of narrow polydispersity at temperatures up to 70°C and possibly higher. This polymerization is thought to be mediated through ylide formation between the NTPP cation and PMMA anion, as supported by ^{1}H NMR assignments using the model compound MIB. The PMMA, NTPP ylide stability is thought to be high, as seen by reaction times of hours as opposed to seconds for anionic MMA polymerizations. The NTPP-mediated polymerizations most likely proceed through free or ion paired enolates that are in rapid equilibrium with the predominant ylides.

This well controlled polymerization technique may allow the synthesis of PMMA polymers and block copolymers at elevated temperatures.

Acknowledgements.

Support for this research by BASF AG, Ludwigshafen, Germany is gratefully acknowledged

References.

(1)   a.Webster, O.W.; Hertler, D.Y.; Sogah, D.Y.; Farnham,W.B.; Babu,T.V.P. *J. Am. Chem.Soc.* **1983,***105*, 5706. b.Wang,J-S.; Jerome, R.;Teyssie, P. *J. Phys. Org. Chem.* **1995**, *8* (4), 208. c. Davis, T.P.; Haddleton, D.M.; Richards, S.N. *J.M.S. Rev Macromol. Chem. Phys.* **1994**, *C34*, 243.

(2)   Baskaran, D. Chakrapani, S.; Sivaram, S.; Hogen-Esch, T.E.; Mueller, A. H. E.; *Macromolecules* **1999,***32*, 2865 and references therein.

(3)   Zagala, A.P.; Hogen-Esch, T.E. *Macromolecules* **1996,** *29*, 3038.

(4)   Baskaran, D.; Mueller, A. H. E.; Kolshorn, H.; Zagala, A.P.; Hogen-Esch, T.E. *Macromolecules* **1997,***30*, 6695.

(5)   Horner, L.; Haufe, J. *Chem. Ber.* **1968**, *101*, 2903

(6)   Zagala, A.P.; Dimov, D.K.; Hogen-Esch, T.E. *Macromol. Symp.* **1998**, *132*, 309.

(7)   Baskaran, D.; Mueller, A. H. E *Macromolecules* **1997**,*30*, 1869.

(8)   Penczek, S.; Kubisa,, P.; Szymanski, R. *Makromol. Chem., Rapid Commun.*, **1991**, *12* , 77.

(9)   Mueller, A. H. E in "Comprehensive Polymer Science".,Vol. 3 J. C. Bevington Ed., Pergamon Press, Oxford 1988.

(10)   Cristau, H.J.; Coste, J.; Truchon, A.; and Cristol, H. *J. Organomet. Chem.*, **1983**, C1-C4 , 241.

(11)   Klippert, H.; Ringsdorf, H. *Makromol. Chem.* **1972**, *153*, 289.

(12)   March,J. "Advanced Organic Chemistry" Fourth Edition, Wiley Interscience , New York 1992.

(13)   Eppley, R. L.; Dixon, J. A. J. Am. Chem. Soc.**1968**, *90*, 1606.

(14)   Weiss, H. Unpublished data.

Macromol. Symp. 157, 183–192 (2000)

Novel Segmented Copolymers by Combination of Controlled Ionic and Radical Polymerizations

Krzysztof Matyjaszewski,* Mircea Teodorescu, Metin H. Acar, Kathryn L. Beers, Simion Coca, Scott G. Gaynor, Peter J. Miller and Hyun-jong Paik

Center for Macromolecular Engineering, Department of Chemistry, Carnegie Mellon University, 4400 Fifth Avenue, Pittsburgh, PA 15213, U.S.A.

Summary. Transformation of different living and non-living polymerization mechanisms to controlled/"living" atom transfer radical polymerization (ATRP) in order to prepare block and graft copolymers is described. The synthesis and characterization of macroinitiators and the resulting segmented copolymers is discussed.

Introduction

Block and graft copolymers represent valuable polymeric materials, as their properties can be designed by a proper choice of the monomers that form the polymeric segments. For example, impact resistant materials or thermoplastic elastomers can be obtained when the constituent blocks of the segmented copolymers are thermodynamically incompatible and microphase separation occurs. Also, amphiphilic copolymers with such applications as hydrogels, stabilizers, surface-modifying agents, compatibilizers in polymer blends, etc. have been prepared[1,2].

Depending on the monomers involved, the segmented copolymers can be synthesized by employing the same polymerization mechanism for all blocks, or by using a combination of different mechanisms. In the former case the synthesis is accomplished by the appropriate sequential addition of the monomers if block copolymers are targeted. In the latter case, the most often used procedure involves the conversion of the functional end groups of the first block (in the case of block copolymers) or the groups distributed along the chain (in the case of graft copolymers) to initiating sites for the polymerization of the second monomer. This approach is employed when the monomers involved do not polymerize by the same mechanism, examples include conversion of vinyl polymerization to ring opening polymerization, cationic to anionic, condensation to radical polymerization, etc.[3,4]

CCC 1022-1360/00/$ 17.50+.50/0

The controlled/"living" polymerizations allow for the preparation of well-defined polymers, i. e. polymers with predetermined molecular weights and low polydispersities, and they have been extensively used for the preparation of segmented copolymers. Until lately, well-defined segmented copolymers could be obtained only by ionic polymerizations. Due to the lack of control of molecular weight and polydispersity, the free-radical methods led to ill-defined polymers, very often accompanied by a certain amount of homopolymer and/or crosslinked material[3,4]. However, recently free-radical techniques have also became available for the preparation of well-defined polymers. The most important types of controlled radical polymerizations (CRP) are: i) stable free-radical polymerization (SFRP), which employs nitroxyl radicals[5]; ii) atom transfer radical polymerization (ATRP), which uses complexes of transition metals in conjunction with alkyl halides[5]; iii) reversible addition-fragmentation chain transfer polymerization (RAFT), which uses dithioesters together with a free-radical initiator[6] or other degenerative transfer processes[7]. So far only SFRP and ATRP have been used to prepare segmented copolymers by combination with other polymerization mechanisms. Table 1 (block copolymers) and Table 2 (graft copolymers) summarize the results published on this subject.

In all the cases presented in Table 1, the block copolymer synthesis consists in at least 2 steps: the macroinitiator synthesis and the polymerization of the second monomer. A particular case is represented by the "one pot" synthesis, when a double-headed initiator is used to simultaneously initiate the formation of both blocks, by two different mechanisms. The method has been used to prepare a polyoxazoline - PSt block copolymer by simultaneous cationic ROP and SFRP[39], and PCL-PSt block copolymers by simultaneous anionic ROP and SFRP or ATRP[40].

In our continuing effort to develop new materials by ATRP, we addressed the field of block and graft copolymers by transformation reactions. The present paper aims to present our work concerning the preparation of segmented copolymers by using controlled/"living" atom transfer radical polymerization in conjunction with another polymerization mechanism, such as anionic, cationic, ROMP, conventional radical or step-growth polymerization. Detailed results are shown in Table 3 (block copolymers) and Table 4 (graft copolymers).

Table 1. Examples of block copolymers prepared by CRP using transformation reactions

Entry	Macroinitiator		Second block		Ref.
	Polymer	Polymn. mechanism	Monomer	Polymn. mechanism	
1	PSt, PBut	anionic	acrylates, St	SFRP	8,9)
2	PEO, PCL, PDMS	anionic ROP	St	SFRP	10-14)
3	PTHF, PCHO	cationic ROP	St	SFRP	15-17)
4	polycarbonate	condensation	St	SFRP	18)
5	PBA, PMMA, PIP	conv. radical	St	SFRP	19)
6	PSt	SFRP	CL	anionic ROP	12)
7	PSt	SFRP	phos + bph, aryleneX$_2$	condensation	18,20)
8	PSt, PEB	anionic	St, 4-acetoxySt MA, BA, MMA	ATRP	21,22)
9	PEO, PCL	anionic ROP	St, MMA, t-BA	ATRP	23-26)
10	PSt, PIb	cationic	St, 4-acetoxySt MA,MMA, IBA	ATRP	27-30)
11	PTHF, PDMS	cationic ROP	St,MA,MMA	ATRP	31,32)
12	PNB, PDCPD	ROMP	St, MA	ATRP	33)
13	PBA, PVDF, PSt, PVAc	conv. radical	St, BA	ATRP	34-36)
14	PMPSil, polyarylene, PSulfone	condensation	St, BA	ATRP	20,37,38)
15	PMMA	ATRP	CL	anionic ROP	12)

PBut - polybutadiene; PSt - polystyrene; PEO - poly(ethylene oxide); PCL - poly(e-caprolactone); PTHF - polytetrahydrofuran; PCHO - poly(cyclohexane oxide); phos - phosgene; bph - bisphenol A; PEB - poly(ethylene-*co*-butylene); PIb - polyisobutylene; PDMS - polydimethylsiloxane; PNB - polynorbornene; PDCPD - polydicyclopentadiene; PVAc - poly(vinyl acetate); PBA - poly(butyl acrylate); PVDF - poly(vinylidene fluoride); PMPSil - poly(methylphenylsilane); PSulfone - polysulfone; PMMA - poly(methyl methacrylate); aryleneX$_2$ - arylene dihalogen; MA - methyl acrylate; IBA - isobornyl acrylate

Table 2. Examples of graft copolymers prepared by CRP using transformation reactions

Entry	Macroinitiator		Second block		Ref.
	Polymer	Polymn. mechanism	Monomer	Polymn. mechanism	
1	PSt	conv. radical	St	SFRP	41)
2	PCL	anionic ROP	MMA	ATRP	42)
3	PIb	cationic	St, IBA	ATRP	43,44)
4	polysiloxane	cationic ROP	St	ATRP	32)
5	PEPDM	coordinative	MMA	ATRP	45)
6	CSPE, PE, PVC	conv. radical	St, MA, BA, MMA	ATRP	46-48)
7	PMPSil	cond.	St	ATRP	49)

PEDM - ethylene-propylene-diene terpolymer; CSPE - chlorosulfonated polyethylene; PE - polyethylene; PVC - poly(vinyl chloride).

Results and discussion

Almost all polymerization mechanisms were combined with ATRP in order to prepare block copolymers. Whenever possible, the first block was prepared in a living manner (Table 3, entries 1-12, 14-17), which allowed a better control of the molecular weights, polydispersities and end functionalities. Thus, "living" anionic polymerization of styrene was terminated with styrene oxide/2-bromoisobutyryl bromide, and after isolation, the polymer was used to initiate ATRP of (meth)acrylates (Table 3, entries 1-3). "Living" cationic polymerizations of styrene and isobutene were also used to prepare ATRP macroinitiators (Table 3, entries 4-9). In the latter case a difunctional polyisobutene terminated with several units of styrene was employed in order to prepare ABA-type block copolymers. In both cases no transformation chemistry was needed because the ATRP initiating sites naturally resulted due to the polymerization characteristics. Mono- (Table 3, entries 10, 11) and difunctional (Table 3, entry 12) PTHF macroinitiators were synthesized by initiating THF polymerization with 2-bromopropionyl bromide/silver triflate or by deactivating a difunctional "living" PTHF with sodium 2-bromopropionate, respectively.

ROMP was also successfully employed to prepare block copolymers in conjunction with ATRP (Table 3, entries 14-17). Polynorbornene and polydicyclopentadiene were synthesized in a "living" manner by using the Schrock catalyst, followed by deactivation with p-(bromomethyl)benzaldehyde to provide a bromobenzyl terminated macroinitiator.

Poly(dimethylsiloxane), commercially available with Si-H end groups and a polysulfone with hydroxy end groups, prepared by the condensation polymerization of bisphenol A and 4,4'-difluorosulfone, were converted to ATRP macroinitiators by reacting the terminal groups with vinylbenzyl chloride in the presence of a Karstedt catalyst or with 2-bromopropionyl bromide, respectively (Table 3, entries13, 19).

Conventional radical polymerization was combined with ATRP in both ways in order to prepare block copolymers. When the first block was prepared by conventional radical polymerization, the macroinitiator resulted directly by using either a thermal initiator containing the ATRP initiating group (Table 3, entry 19) or CCl_4 as a telogen (Table 3, entry 18). The other way involves the preparation of the first block by ATRP initiated by an initiator containing an thermolabile azo group, which can be decomposed later, at a higher temperature, to initiate the polymerization of a second monomer by conventional radical polymerization (Table 3, entry 22). Because of the lack of control in the conventional radical polymerization step, the synthesized block copolymers had higher polydispersities.

Commercially available polymers were used as macroinitiators or precursors in almost all graft copolymer syntheses. Thus, brominated poly(isobutene-co-p-methylstyrene) with 1.2 mole-% bromobenzyl groups (PIb-Br), chlorosulfonated polyethylene with sulfonyl chloride groups as ATRP initiating sites and a statistical copolymer of vinyl chloride with 1 mole-% vinyl chloroacetate (PVCA) have been used as macroinitiators without any further transformation (Table 4). In other cases, transformation reactions were carried out in order to convert the functional groups of the precursor into initiating sites for ATRP. Thus, the vinyl groups of poly(dimethyl-co-vinylmethylsiloxane) (PSilox) were reacted with 2-(4-chloromethylphenyl)ethyldimethylsilane to produce a polymer with pendant benzyl chloride functionalities, which was used to initiate ATRP of styrene (Table 4, entry 3). Similarly, commercially available poly(ethylene-co-glycidyl methacrylate) (PEGM) was converted to a macroinitiator by reacting the epoxy groups with chloroacetic acid in the presence of tetrabutylammonium hydroxide in xylene at 115°C (Table 4, entry 6).

Table 3. Block copolymers by combining different polymerization mechanisms with ATRP

Entry	Macroinitiator				Second monom.	Block copolymer		Ref.
	Polymer	Polymn. mechanism	M_n	M_w/M_n		M_n	M_w/M_n	
1	PSt	anionic	12200	1.04	MA	19000	1.07	22)
2	"	"	"	"	BA	23300	1.14	"
3	"	"	"	"	MMA	25000	1.36	"
4	PSt	cationic	2100	1.17	MA	6300	1.20	27)
5	"	"	"	"	MMA	11100	1.57	"
6	PIb	cationic	7800	1.31	St	13400	1.18	28)
7	"	"	"	"	MA	12200	1.41	"
8	'	"	"	"	MMA	22500	1.45	"
9	"	"	"	"	IBA	18900	1.44	"
10	PTHF	cationic ROP	15400	1.39	St	31000	1.46	31)
11	"	"	"	"	MA	28500	1.32	"
12	"	"	20000	1.71	MMA	71000	1.34	"
13	PDMS	"	9800	2.40	St	20700	1.60	32)
14	PNB	ROMP	5700	1.21	St	26000	1.16	33)
15	"	"	"	"	MA	12000	1.31	"
16	PDCPD	"	7000	1.24	St	17000	1.37	"
17	"	"	"	"	MA	21000	1.47	"
18	PVAc	radical	3600	1.81	St	24300	1.42	36)
19	PSt	"	39400	2.48	BA	127000	1.49	"
20	PSulfone	condensation	4030	1.5	St	10700	1.1	38)
21	"	"	"	"	BA	15300	1.2	"
22	PBA	ATRP	7500	1.15	VAc*	41800	3.56	36)

*polymerized by conventional radical polymerization.

Table 4. Graft copolymers by combining different polymerization mechanisms with ATRP

No	Macroinitiator					Mon.	Graft copolymer				Ref
	Polym	Polymn mech.	M_n (10^{-3})	$M_w/$ M_n	T_g /°C		Mon. /wt-%	M_n (10^{-3})	$M_w/$ M_n	T_g /°C	
1	PIb-Br	cationic	108	2.3	-60	St	69	250	2.4	-60/98	43)
2	"	"	"	"	"	IBA	21	181	2.5	-10	"
3	PSilox	CROP	6.6	1.8	-	St	54	14.8	2.1	-	32)
4	CSPE	radical	14.9	2.32	-15	St	-	85.6	1.78	-10/87	46)
5	"	"	"	"	"	MMA	-	26.3	1.75	-2	"
6	PEGM	"	-	-	-15	St	69	-	-	-15/ 108	48)
7	PVCA	"	47.4	2.66	83	St	80	99.5	3.72	80	47)
8	"	"	"	"	"	MA	50	57.7	2.40	21	"
9	"	"	"	"	"	MMA	60	83.6	4.94	111	"
10	"	"	"	"	"	BA	65	81.4	2.44	-19	"
11	PBPEA	"	27.3	2.3	-	BA	-	1,500	1.5	-	-

A particular case of graft copolymers is represented by densely grafted or "brush" copolymers. They were synthesized using macroinitiators with ATRP initiating sites at each repeating unit along the backbone, prepared either by free radical polymerization of 2-(2-bromopropionoxy)ethyl acrylate (BPEA) (Table 4, entry 11) or by esterification with 2-bromoisobutyryl bromide of well-defined poly(2-trimethylsilyloxyethyl methacrylate), which has already been prepared by ATRP (Scheme 1)[50].

Conclusions

Atom transfer radical polymerization (ATRP) has been successfully used in combination with other living and non-living polymerization mechanisms to prepare block and graft copolymers.

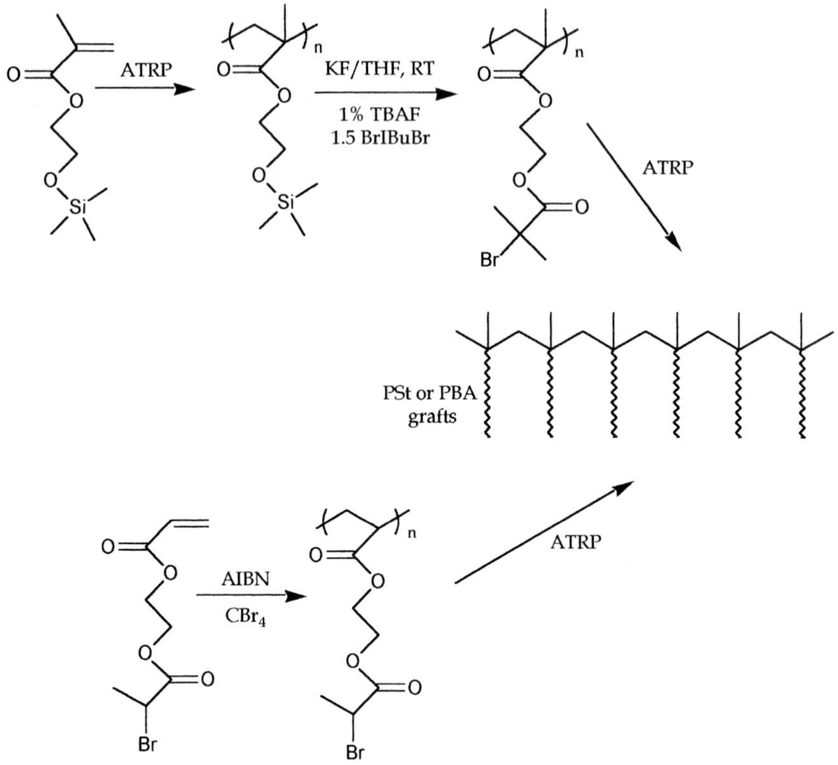

Scheme 1

Acknowledgment

Financial support form the Industrial Members of the ATRP Consortium at CMU is greatly appreciated.

References

1. B. Ameduri, B. Boutevin, Ph. Gramain, *Adv. Polym. Sci.* **127,** 87 (1997)
2. M. Pitsikalis, S. Pispas, J. W. Mays, N. Hadjichristidis, *Adv. Polym. Sci.* **135,** 1 (1998)
3. F. Schue, in: *Comprehensive Polymer Science*, G. Allen, J. C. Bevington (Eds.), Pergamon Press, Oxford 1989, Vol. 6, p. 359
4. P. F. Rempp, P. J. Lutz, in: *Comprehensive Polymer Science*, G. Allen, J. C. Bevington (Eds.), Pergamon Press, Oxford 1989, Vol. 6, p. 403
5. K. Matyjaszewski (Ed.), *ACS Symposium Series 685: Controlled Radical Polymerization*, American Chemical Society, Washington, D.C., 1998
6. J. Chiefari, Y. K. Chong, F. Ercole, J. Krstina, J. Jeffery, T. P. T. Le, R. T. A. Mayadunne, G. F. Meijs, K. L. Moad, E. Rizzardo, S. H. Thang, *Macromolecules* **31,** 5559 (1998)

7. K. Matyjaszewski, S. Gaynor, J.-S. Wang, *Macromolecules* **28,** 2093 (1995)
8. E. Yoshida, T. Ishizone, A. Hirao, S. Nakahama, T. Takata, T. Endo, *Macromolecules* **27,** 3119 (1994)
9. S. Kobatake, H. J. Harwood, R. P. Quirk, D. B. Priddy, *Macromolecules* **31,** 3735 (1998)
10. E. Yoshida, S. Tanimoto, *Macromolecules* **30,** 4018 (1997)
11. X. Chen, B. Gao, J. Kops, W. Batsberg, *Polymer* **39,** 911 (1998)
12. C. J. Hawker, J. L. Hedrick, E. E. Malmstrom, M. Trollsas, D. Mecerreyes, G. Moineau, P. Dubois, R. Jerome, *Macromolecules* **31,** 213 (1998)
13. E. Yoshida, Y. Osagawa, *Macromolecules* **31,** 1446 (1998)
14. Y. Wang, S. Chen, J. Huang, *Macromolecules* **32,** 2480 (1999)
15. E. Yoshida, A. Sugita, *Macromolecules* **29,** 6422 (1996)
16. E. Yoshida, A. Sugita, *J. Polym. Sci.* **36,** 2059 (1998)
17. T. G. Yildirim, Y. Hepuzer, G. Hizal, Y. Yagci, *Polymer* **40,** 3885(1999)
18. I. Q. Li, D. M. Knauss, Y. Gong, B. Pan, B. A. Howell, D. B. Priddy, *Am. Chem. Soc., Polym. Prepr.* **39(2),** 598 (1998)
19. I. Q. Li, B. A. Howell, M. T. Dineen, P. E. Kastl, J. W. Lyons, D. M. Meunier, P. B. Smith, D. B. Priddy, *Macromlecules* **30,** 5195(1997)
20. R. D. Miller, C. R. Hawker, J. L. Hedrick, M. Trollsas, M. Husemann, A. Heise, G. Klaerner, *PMSE* **80,** 24 (1999)
21. K. Jankova, J. Kops, X. Chen, W. Batsberg, *Macromol. Rapid. Commun.* **20,** 219 (1999)
22. M. H. Acar, K. Matyjaszewski, *Macromol. Chem. Phys.* **200,** 1094 (1999)
23. K. Jankova, X. Chen, J. Kops, W. Batsberg, *Macromolecules* **31,** 538 (1998)
24. K. Jankova, J. H. Truelsen, X. Chen, J. Kops, W. Batsberg, *Polym. Bull.* **42,** 153 (1999)
25. B. Reining, H. Keul, H. Hocker, *Polymer* **40,** 3555 (1999)
26. M. Bednarek, T. Biedron, P. Kubisa, *Macromol. Rapid Commun.* **20,** 59 (1999)
27. S. Coca, K. Matyjaszewski, *Macromolecules* **30,** 2808 (1997)
28. S. Coca, K. Matyjaszewski, *J. Polym. Sci.: Part A: Polym. Chem.* **35,** 3595 (1997)
29. X. Chen, B. Ivan, J. Kops, W. Batsberg, *Macromol. Rapid Commun.* **19,** 585 (1998)
30. K. Jankova, J. Kops, X. Chen, B. Gao, W. Batsberg, *Polym. Bull.* **41,** 639 (1998)
31. A. Kajiwara, K. Matyjaszewski, *Macromolecules* **31,** 3489 (1998)
32. Y. Nakagawa, P. J. Miller, K. Matyjaszewski, *Polymer* **39,** 5163 (1998)
33. S. Coca, H.-j. Paik, K. Matyjaszewski, *Macromolecules* **30,** 6513 (1997)
34. M. Destarac, B. Boutevin, *Am. Chem. Soc., Polym. Prepr.* **39(2),** 568 (1998)
35. Z. Zhang, S. Ying, Z. Shi, *Polymer* **40,** 1341 (1999)
36. H.-j. Paik, M. Teodorescu, J. Xia, K. Matyjaszewski, *Macromolecules* submitted
37. L. Lutsen, G. P.-G. Cordina, R. G. Jones, F. Schue, *Eur. Polym. J.* **34,** 1829 (1998)
38. S. G. Gaynor, K. Matyjaszewski, *Macromolecules* **30,** 4241 (1997)
39. M. W. Weimer, O. A. Scherman, D. Y. Sogah, *Macromolecules* **31,** 8425(1998)
40. D. Mecerreyes, G. Moineau, P. Dubois, R. Jerome, J. L. Hedrick, C. J. Hawker, E. E. Malmstrom, M. Trollsas, *Angew. Chem. Ind. Ed.* **37,** 1274 (1998)
41. C. J. Hawker, *Angew. Chem. Int. Ed. Engl.* **34,** 1456 (1995)
42. J. L. Hedrick, B. Atthoff, K. A. Boduch, C. J. Hawker, D. Mecerreyes, R. D. Miller, M. Trollsas, *PMSE* **80,** 104 (1999)
43. S. G. Gaynor, K. Matyjaszewski, in: *ACS Symposium Series 685: Controlled Radical Polymerization,* K. Matyjaszewski (Ed.), American Chemical Society, Washington, D.C. 1998, p. 396
44. T. Fonagy, B. Ivan, M. Szesztay, *Macromol. Rapid Commun.* **19,** 479 (1998)
45. X. Wang, N. Luo, S. Ying, *Polymer* **40,** 4515 (1999)
46. WO 98/40415 (1998), invs.: S. Coca, S. G. Gaynor, K. Matyjaszewski,

47. H.-J. Paik, S. G. Gaynor, K. Matyjaszewski, *Macromol. Rapid Commun.* **19,** 47 (1998)
48. P. J. Miller, M. Teodorescu, M. L. Peterson, K. Matyjaszewski, *Am. Chem. Soc., Polym. Prepr.* **40(2),** 00 (1999)
49. R. G. Jones, S. J. Holder, *Macromol. Chem. Phys.* **198,** 3571 (1997)
50. K. L. Beers, S. G. Gaynor, K. Matyjaszewski, S. S. Sheiko, M. Moeller, *Macromolecules* **31,** 9413 (1998)

Ru(II)-Mediated Living Radical Polymerization:

Block and Random Copolymerizations of

N,N-Dimethylacrylamide and Methyl Methacrylate

Masahide Senoo, Yuzo Kotani, Masami Kamigaito, Mitsuo Sawamoto*

Department of Polymer Chemistry, Graduate School of Engineering,
Kyoto University, Kyoto 606-8501, Japan

SUMMARY: $RuCl_2(PPh_3)_3$ led to living radical copolymerization of *N,N*-dimethylacrylamide (DMAA) and methyl methacrylate (MMA) in conjunction with a halide-initiator (R–X; $CHCl_2COPh$, CCl_3Br) and $Al(Oi\text{-}Pr)_3$ in toluene at 80°C. Both the monomers were polymerized at almost the same rate into random copolymers, where the number-average molecular weights (M_n) increased in direct proportion to weight of the obtained polymers, and the molecular weight distributions (MWDs) were narrow throughout the reactions (M_w/M_n = 1.2–1.6). MMA was consumed faster in the copolymerization than in the homopolymerization, which was due to the interaction of DMAA with the ruthenium complex. The Ru(II)-based initiating system was also effective in block copolymerization of DMAA and MMA.

Introduction

Extensive efforts have been directed towards developing living polymerizations[1]. Although it is increasingly important to develop such precision processes applicable to polar functional monomers, not many living polymerizations are tolerant to polar groups. One of the most significant features of radical polymerizations is their versatility, as indicated by the wide variety of applicable monomers and their facile copolymerizability, in contrast to ionic counterparts, where polar functional groups often deactivate ionic growing species and the monomer reactivity greatly changes with electronic nature of their substituents.

We have developed living radical polymerizations of methacrylates and styrenes catalyzed by transition metal complexes such as ruthenium[3], iron[4], nickel[5], and rhenium[6] (Eq. 1). These living polymerizations proceed via the metal-assisted reversible activation of carbon–halogen terminals derived from a halide initiator, where the metal center undergoes one-electron redox reaction. A key to these living polymerizations is an equilibrium between the dormant and the active species, where the equilibrium is shifted to the former[2f, 2g]. This methodology has proved effective because the growing radical species is kept at low concentrations to diminish bimolecular termination reactions[2], the most serious chain breaking processes in free radical polymerization.

$$(1)$$

More recently, we have found that a ruthenium(II)-complex, $RuCl_2(PPh_3)_3$, induced living radical polymerization of a polar monomer, N,N-dimethylacrylamide (DMAA), in conjunction with a halide initiator (R–X) and $Al(Oi\text{-}Pr)_3$ to give polymers of controlled molecular weights and narrow molecular weight distributions (MWDs)[7]. This study deals with the living radical, statistic (or random) copolymerization of DMAA and methyl methacrylate (MMA) with the Ru(II)-based initiating system and the synthesis of their block copolymers by sequential living polymerization.

Living Random Copolymerization of MMA and DMAA

An equimolar mixture of MMA and DMAA was polymerized with $CHCl_2COPh/$ $RuCl_2(PPh_3)_3/Al(Oi\text{-}Pr)_3$ in toluene at 80°C. As shown in Fig. 1, both monomers were polymerized almost simultaneously. MMA was consumed much faster than in the homopolymerization where it takes over 4 days to reach 90% conversion under the same conditions[8], while such a dramatic acceleration was not seen for DMMA in its homo- and copolymerizations. The fast consumption of MMA is due to some interaction between $RuCl_2(PPh_3)_3$ and the amide group of DMAA or of the DMAA unit in the polymers, similarly to added amines, which also accelerate the Ru(II)-catalyzed polymerization of MMA[9]. A bromide-initiator, CCl_3Br, also induced random copolymerization of the two monomers in conjunction with $RuCl_2(PPh_3)_3$ and $Al(Oi\text{-}Pr)_3$.

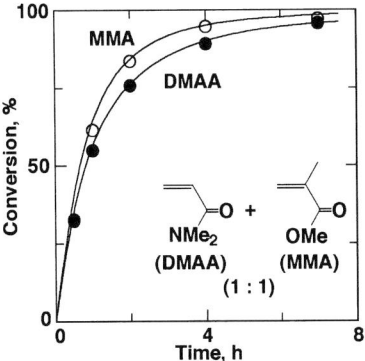

Fig. 1: Polymerization of mixture of DMAA (●) and MMA (○) with CHCl₂COPh/
RuCl₂(PPh₃)₃/Al(Oi-Pr)₃ in toluene at 80°C: [DMAA]₀ = [MMA]₀ = 1.0 M; [CHCl₂COPh]₀ =
20 mM; [RuCl₂(PPh₃)₃]₀ = 10 mM; [Al(Oi-Pr)₃]₀ = 40 mM.

Fig. 2 plots the number-average molecular weights (M_n) and MWDs of the polymers. Irre-
spective of the halogens in the initiators, the size-exclusion chromatograph (SEC) curves
shifted to higher molecular weights as the reaction proceeds, keeping unimodal and narrow
MWDs. With CCl₃Br, the M_n increased with polymer yield and was very close to the calculated value assuming that one molecule of the initiator generates one living polymer chain.
The M_n were higher with CHCl₂COPh similarly to the homopolymerization[7]. These are due
to the low reactivity of the C–Cl bonds relative to C–Br, which results in slower initiation as
well as slower interconversion between the dormant and the radical species (cf. Eq. 1).

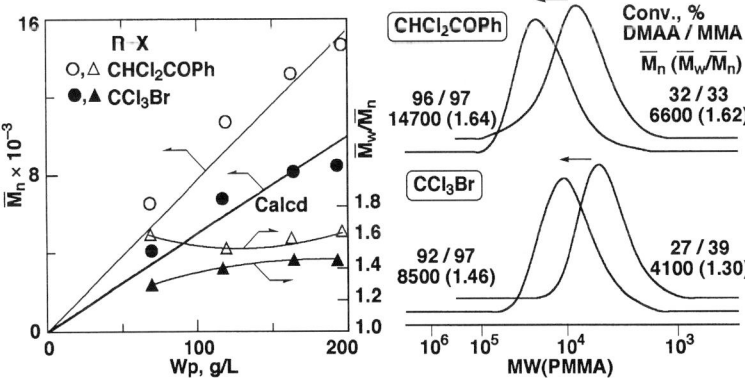

Fig. 2: M_n, M_w/M_n, and SEC curves of copolymers of DMAA and MMA obtained with R–X/
RuCl₂(PPh₃)₃/Al(Oi-Pr)₃ in toluene at 80°C: [DMAA]₀ = [MMA]₀ = 1.0 M; [R–X]₀ = 20 mM;
[RuCl₂(PPh₃)₃]₀ = 10 mM; [Al(Oi-Pr)₃]₀ = 40 mM. R–X: (○, △) CHCl₂COPh; (●, ▲)
CCl₃Br.

Another copolymerization at a feed ratio DMAA/MMA = 1/9 also gave similar results; both monomers were polymerized almost simultaneously to give polymers whose molecular weights increased with conversion (Fig. 3).   The MWDs were unimodal throughout the reactions and narrower than those obtained in the copolymerization of the equimolar mixture. This is due to the lower content of DMAA, which results in broader MWDs when polymerized with the Ru(II)-based system.

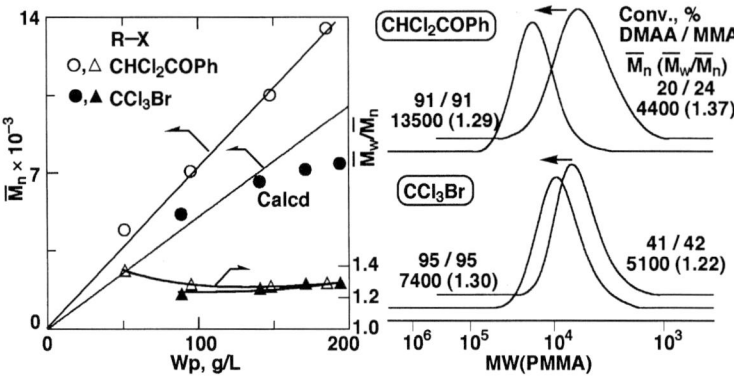

Fig. 3: M_n, M_w/M_n, and SEC curves of copolymers of DMAA and MMA obtained with R–X/RuCl$_2$(PPh$_3$)$_3$/Al(Oi-Pr)$_3$ in toluene at 80°C: [DMAA]$_0$ = 0.20 M; [MMA]$_0$ = 1.8 M; [R–X]$_0$ = 20 mM; [RuCl$_2$(PPh$_3$)$_3$]$_0$ = 10 mM; [Al(Oi-Pr)$_3$]$_0$ = 40 mM.   R–X: (O, △) CHCl$_2$COPh; (●, ▲) CCl$_3$Br.

The ^{1}H NMR spectrum (Fig. 4C) of the products obtained from the equimolar mixture is different from a simple sum of those of homopolymers (Fig. 4A and 4B).   The signals (d) of the methyl ester protons of MMA units in Fig. 4C are not single, unlike that of the homopoly(MMA) (Fig. 4B), and the methyl protons of DMAA units (signal a) were broadened in comparison with the homoply(DMAA) (Fig. 4A).   Similar changes were also observed for other absorptions.   These show that the products are not mixtures of the homopolymers but statistical or random copolymers.   This is also the case for the products from the 1/9-mixture of DMAA and MMA (Fig. 4D).   The number-average degrees of polymerization (DP$_n$) of each monomer, calculated from the peak intensity ratios of the main-chain units to the phenyl groups originated from CHCl$_2$COPh, were slightly higher than the calculated values based on the feed ratio and the gas-chromatographic conversions.   However, the unit ratios were close to the calculated values.

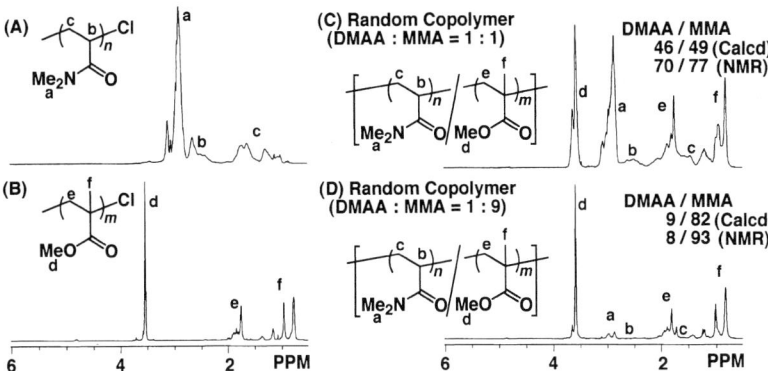

Fig. 4: ¹H NMR spectra of poly(DMAA) (A), poly(MMA) (B), copolymers of DMAA and MMA at 1/1 (C) and 1/9 (D) feed ratio obtained with $CHCl_2COPh/RuCl_2(PPh_3)_3/Al(Oi\text{-}Pr)_3$ in toluene at 80°C.

These results indicate that the $R–X/RuCl_2(PPh_3)_3/Al(Oi\text{-}Pr)_3$ initiating system induced living random copolymerization of DMAA and MMA.

Acceleration of Polymerization by DMAA

As described above, the consumption of MMA was much faster than that in the homopolymerization. This indicates a fast cross propagation between the two monomers or due to some interaction of DMAA with the Ru(II)-catalyst, which increases the catalytic activity.

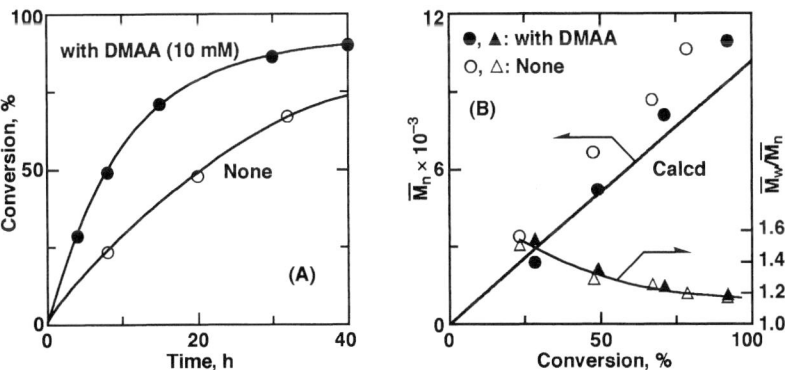

Fig. 5: Polymerization of MMA with $CHCl_2COPh/RuCl_2(PPh_3)_3/Al(Oi\text{-}Pr)_3$ in toluene at 80°C in the presence (●) and the absence (○) of DMAA: $[MMA]_0 = 2.0$ M; $[CHCl_2COPh]_0 = 20$ mM; $[RuCl_2(PPh_3)_3]_0 = 10$ mM; $[Al(Oi\text{-}Pr)_3]_0 = 40$ mM; $[DMAA]_0 = 10$ mM.

To investigate this, homopolymerization of MMA was carried out in the presence of small amount (10 mM) of DMAA, equimolar to $RuCl_2(PPh_3)_3$ or only 0.5 mol% of MMA. As shown in Fig. 5A, a small amount of DMAA accelerated the MMA polymerization almost 2.5 times. Such a large increase in rate by a small amount of added DMAA indicates that it is mainly due to some interaction between DMAA or the repeat unit in the polymer chain and the Ru(II) complex. Interestingly, a similar acceleration also occurs on addition of nitrogen compounds like butyl amines[9]. The M_n increased in direct proportion to monomer conversion and agreed well with the calculated, as with that obtained without DMAA (Fig. 5B).

Block Copolymers of MMA and DMAA

We also investigated the synthesis of AB-type block copolymers of MMA and DMAA with the Ru(II)-based initiating system. MMA was thus polymerized with a monofunctional chloride initiator, H–$(MMA)_2$–$Cl^{10)}$, coupled with $RuCl_2(PPh_3)_3$ and $Al(Oi\text{-}Pr)_3$ to give living homopoly(MMA) with M_n = 11000 and M_w/M_n = 1.32, into which was added an equimolar amount of DMAA (Fig. 6). Added DMAA was smoothly polymerized to increase the molecular weights with keeping monomodal MWDs (M_n = 23900, M_w/M_n = 1.80). This shows the formation of block copolymers of MMA and DMAA.

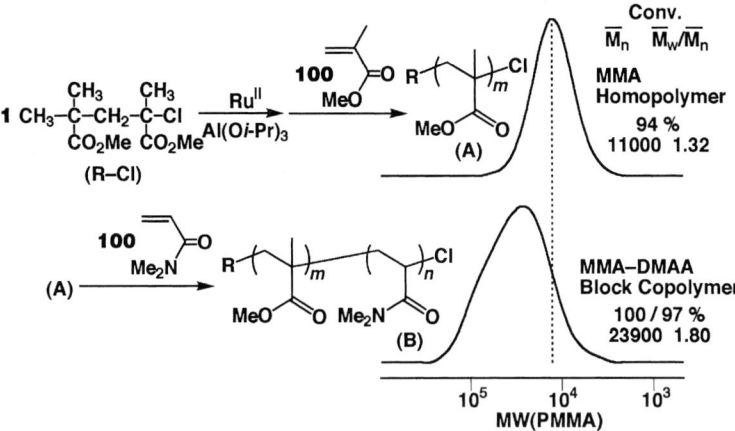

Fig. 6: SEC curves of poly(MMA) (A) and MMA–DMAA block copolymers (B) obtained with 1/$RuCl_2(PPh_3)_3$/$Al(Oi\text{-}Pr)_3$ in toluene at 80°C: $[MMA]_0$ = 2.0 M; $[\mathbf{1}]_0$ = 20 mM; $[RuCl_2(PPh_3)_3]_0$ = 10 mM; $[Al(Oi\text{-}Pr)_3]_0$ = 40 mM; DMAA/MMA = 1/1.

Fig. 7 also supports the formation of block copolymers. As shown in Fig. 7B, the spectrum showed absorptions (d–f) due to DMMA repeat units in addition to those of MMA units. The absorptions of MMA units remain sharp, unlike the spectra of the random copolymers (cf. Fig. 4C). The methyl ester protons adjacent to the PMMA ω-end chloride (a') completely disappeared in Fig. 4D, which suggests that the polymerization of DMAA was initiated from the living PMMA to form AB-type block copolymers. The unit ratio of MMA to DMAA measured by their peak intensity ratios was 0.90 close to that from the calculated value (1.02). The copolymer was soluble in methanol whereas the prepolymer of MMA was insoluble. These results indicate that the R–Cl/RuCl$_2$(PPh$_3$)$_3$/Al(Oi-Pr)$_3$ initiating system is effective in the synthesis of block copolymers of DMAA and MMA.

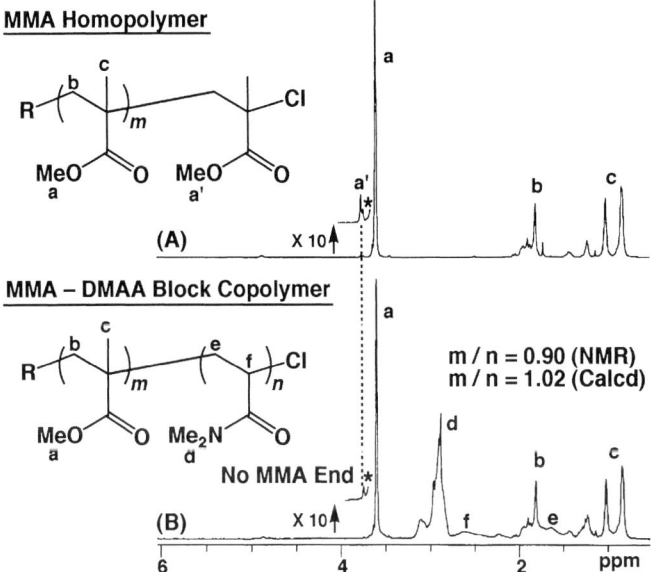

Fig. 7: ^{1}H NMR spectra of poly(MMA) (A) and MMA–DMAA block copolymers (B) ob-tained with **1**/RuCl$_2$(PPh$_3$)$_3$/Al(Oi-Pr)$_3$ in toluene at 80°C. The lines marked by * are as-cribed to satellite lines.

Acknowledgment. With appreciation M.S. and M.K. acknowledge the support from the New Energy and Industrial Technology Development Organization (NEDO) under the Ministry of International Trade and Industry (MITI), Japan, through the grant for "Precision Catalytic Polymerization" in the Project "Technol-ogy for Novel High-Functional Material" (1996–2000).

References

1. For recent reviews on precision control in polymerization, see: (a) *Catalysis in Precision Polymerization*, S. Kobayashi (Ed.), Wiley, Chichester 1997. (b) *New Methods of Polymer Synthesis*; J. R. Ebdon and G. C. Eastmond (Eds.), Blackie, Glasgow 1995, Vol. 2. (c) T. Aida, *Prog. Polym. Sci.* **19**, 469 (1994).
2. For recent reviews on living radical polymerizations, see: (a) M. K. Georges, R. P. N. Veregin, P. M. Kazmaier, G. K. Hamer, *Trends Polym. Sci.* **2**, 66 (1994). (b) T. P. Davis, D. Kukulj, D. M. Haddleton, D. R. Maloney, *Trends Polym. Sci.* **3**, 365 (1995). (c) E. E. Malmström, C. J. Hawker, *Macromol. Chem. Phys.* **199**, 823 (1998). (d) M. Sawamoto, M. Kamigaito, *Trends Polym. Sci.* **4**, 371 (1996). (e) D. Colombani, *Prog. Polym. Sci.* **22**, 1649 (1997). (f) *Controlled Radical Polymerization*, K. Matyjaszewski (Ed.), ACS Symposium Series 685, American Chemical Society, Washington DC, 1998. (g) M. Sawamoto, M. Kamigaito, in: *Synthesis of Polymers (Materials Science and Technology Series)*, A.-D. Schlüter (Ed.), Wiley-VCH, Weinheim 1999, Chapter 6. (h) M. Sawamoto, M. Kamigaito, *CHEMTECH* **29**(6), 30 (1999).
3. M. Kato, M. Kamigaito, M. Sawamoto, T. Higashimura, *Macromolecules* **28**, 1721 (1995).
4. T. Ando, M. Kamigaito, M. Sawamoto, *Macromolecules* **30**, 4507 (1997).
5. H. Uegaki, Y. Kotani, M. Kamigaito, M. Sawamoto, *Macromolecules* **30**, 2249 (1997).
6. Y. Kotani, M. Kamigaito, M. Sawamoto, *Macromolecules* **32**, 2420 (1999).
7. M. Senoo, Y. Kotani, M. Kamigaito, M. Sawamoto, *Macromolecules*, submitted.
8. T. Ando, M. Kato, M. Kamigaito, M. Sawamoto, *Macromolecules* **29**, 1070 (1996).
9. S. Hamasaki, M. Kamigaito, M. Sawamoto, *Polym. Prepr. Jpn.* **48**(2), 134 (1999).
10. T. Ando, M. Kamigaito, M. Sawamoto, *Macromolecules* **31**, 6708 (1998).

Novel Polymers from Atom Transfer Polymerisation
Mediated by Copper(I) Schiff Base Complexes

David M. Haddleton*, Alex M. Heming, Adam P. Jarvis, Afzal Khan, Andrew

Marsh, Sebastien Perrier, Stefan A. F. Bon, Stuart G. Jackson, Ryan Edmonds,

Elizabeth Kelly, Dax Kukulj and Carl Waterson.

Department of Chemistry, University of Warwick, Coventry, CV4 7AL, UK

D.M.Haddleton@warwick.ac.uk

SUMMARY: The use of copper(I) Schiff base complex catalysed atom transfer polymerisation of methacrylates is described. The use of a range of functional and multi-functional initiators enables the synthesis of a range of functional and star polymers to be prepared under undemanding synthetic conditions. End capping with silyl enol ethers allows for ω-functional polymers. The combination of novel initiators, functional monomers and end capping allows an unprecedented array of macromolecular structures to be produced with limited need for protecting group chemistry.

Introduction

Transition metal mediated living polymerisation has emerged as an efficacious method for

the living polymerization of vinyl monomers since its inception by Sawamoto using Ru(II)

catalysts [1] [2] and Matyjaszewski with Cu(I) bpy complexes [3]. This chemistry allows the

synthesis of a wide range of novel polymers under relatively undemanding conditions. The

process is inert to most functional groups and impurities present in monomers, reagents and

solvents [4]. Indeed one of the only conditions that seems to destroy the catalyst is low pH,

especially where may become protonated thus changing the nature of the catalyst [5]. An

impressive range of new polymers has been reported, for example, Fukuda and co-workers [6]

have synthesised glycopolymers by metal mediated living polymerisation. As an example of

the ability to prepare macromolecules with controlled architecture the synthesis of block copolymers by transition metal mediated living polymerisation has emerged by two synthetic routes;

1.   Synthesis of a well defined A block by living radical polymerisation that is isolated and subsequently used as a macroinitiator for the reinitiation of a second monomer.

2.   Synthesis of a macroinitiator by introduction of initiating groups onto polymers by chemical modification of a pre-formed polymer e.g. from condensation or ring-opening polymerisation.

We have been using of a range of Schiff base ligands used in conjunction with Cu(I)Br and an appropriate initiator as a versatile and extremely effective living polymerisation system for acrylics and other vinyl monomers [7]. In this paper we illustrate some of the recent developments that we have been working on and we report a range of polymers that have been synthesised for diverse applications. The examples chosen serve to illustrate the diversity of this chemistry.

Experimental

General Information. For general procedures see previous publications [7] [8]. All reactions were carried out using standard Schlenk techniques under a nitrogen atmosphere. Methyl methacrylate and styrene were purified by passing down an activated basic alumina column so as to remove inhibitor, water and other protic impurities. For detailed procedures the reader is directed to both our previous publications in the area and future detailed papers in each area [7] [8].

Typical synthesis of a macroinitiator. Synthesis of Kraton L-1203 Macroinitiator. Kraton L-1203 168.15 g (0.04 mol) was dissolved in anhydrous tetrahydrofuran 600 mL. Triethylamine 8.4 mL (0.06 mol) was added to the mixture followed by the addition with stirring of 2-bromo-2-methyl propionyl bromide 7.4 mL (0.06 mol). The reaction was allowed to stir overnight at room temperature. The viscous product was dissolved in $CHCl_3$ 500 mL and the solution was sequentially washed with saturated $NaHCO_3$ solution and water. The $CHCl_3$ layer was dried with $MgSO_4$ filtered and the solvent was removed to leaving a clear colorless viscous liquid. Yield = 165.9 g

Typical polymerisation with a macroinitiator: Polymerisation of MMA with 1 as initiator;
[MMA]/[I]/[Cu]/[L] = 100/1/1/2 in 66% toluene solution. Initiator **1**, (0.268 g, 0.5 mmol)
Cu(I)Br (0.072 g, 0.5 mmol) were dissolved in deoxygenated toluene (10.6 mL) in a Schlenk
tube prior to the addition of *N*-propyl-2-pyridinalmethanime (0.180 g, 1.15 mmol). The
solution was further deoxygenated via three freeze pump thaw cycles and the solution heated
to 90 °C. Deoxygenated inhibitor free MMA (10.6 mL) was subsequently added (t = 0).
Samples were removed periodically for analysis, via syringe. Polymerisation of other
monomers with **1** as initiator was carried out under similar conditions.

General instrumentation. GPC was carried out using a Polymer Laboratories (PL) guard
column (50 × 7.5 mm), and two Mixed-C columns (300 × 7.5 mm). THF was used as the
eluent at a flow rate of 1 mL/min, and data were collected at 1 point/s from a DRI detector.
The system was calibrated with log molecular weight expressed as a third order polynomial
of elution volume based on Polymer Laboratories PMMA and PSTY standards and pure
samples of MMA dimer and trimer. The SEC data are presented as 'PMMA equivalent' or
'PSTY equivalent' molecular weights. Elemental analysis was performed using a CE440
Elemental Analyser, Leeman Labs Inc. Infrared analysis were carried out with a Bruker
Vector 22 FTIR spectrometer equipped with a Golden Gate Single Reflection Diamond ATR
accessory, P/N 10500 series, Graseby Specac. NMR spectra were recorded on Bruker AC250
and AC400 spectrometers.

Results and Discussion

Polymerisation is achieved by using α-activated bromo initiators, **1**, in conjunction with
copper(I) Schiff base, **2**, ligands. Ligands of type **2** are synthesized from the condensation of
pyridine 2-carbaldehyde with primary amines in quantitative yield [7]. The abundance of
available primary amines lead to the availability of a range of catalysts with varying
solubility and redox potential.

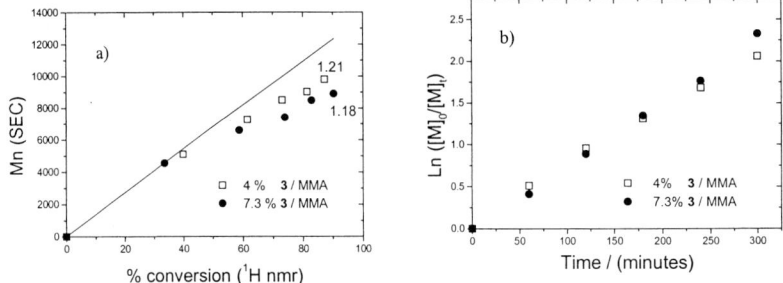

Polymerisation of functional monomers.

Many different monomers, including both hydrophobic and hydrophilic monomers may be polymerised with bromo initiators in conjunction with copper(I) based catalysts. For

Fig. 1: Statistical copolymers of **3** and MMA with **1** as initiator and Cu(I)Br/n-propyl **1** catalyst a) Evolution of Mn with conversion and b) first order kinetic plots

example, hydrophobic fluorine containing monomers, **3**, may be statistically co-polymerized with MMA to give polymers with narrow PDi, controlled Mn in a living polymerisation, Fig. 1. Blends of these statistical co-polymers with MMA form coatings with fluorine rich surfaces as seen by contact angle measurements with water; 4% **3** 104.5^0 (adv.), 74.7^0 (rec.); 7.4 % **3** 104.5^0 (adv), 74.7^0 (rec.); PMMA 72.8^0 (adv.), 59.2^0 (rec.).

Use of functional initiators to α-functional polymers

Functional initiators may be easily prepared by the condensation of both aromatic and aliphatic alcohols with 2-bromo isobutyrate, as above. For example, a range of phenols with methoxy, **4**, primary amino, **5**, and aldehyde, **6**, functional initiators are effective initiators for the living polymerisation of MMA leading to polymers with corresponding α-functionality,

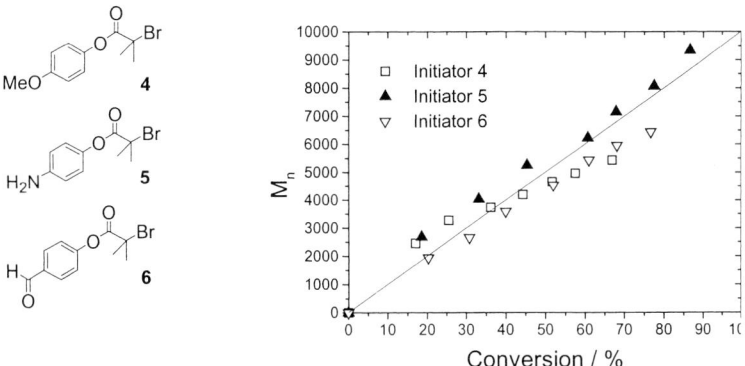

Fig. 2: Evolution of Mn with conversion for the polymerisation of MMA with phenolic ester functional initiators, **4-6**, under standard polymerisation conditions.

Fig 2. This approach can be utilised to prepare a wide-range of α-functional narrow molecular mass distribution polymers of varying composition

Star polymers based on simple sugars.

An extension of the above approach is the preparation of multi-functional initiators from polyols. This has been previously exploited by both Gnanou [9] and Sawamoto [10] for the synthesis of star polymers from calixarenes. We have been using this approach with simple sugars and have previously reported the use of glucose to produce 5-arm star polymers. This is illustrated by the use of sucrose that forms an octa-functional initiator on reaction with eight equivalents of 2-bromo isobutyrate.

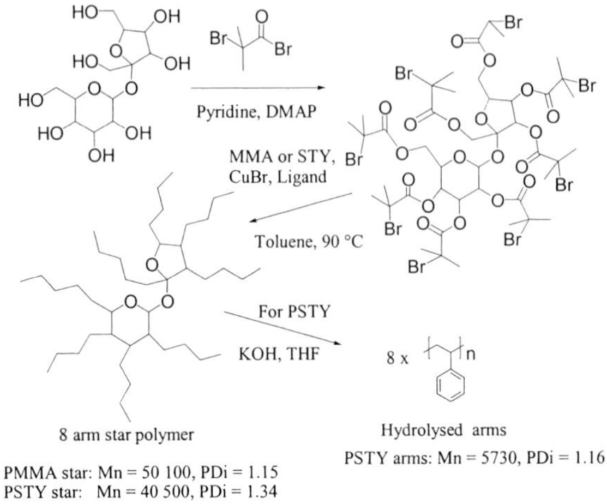

206

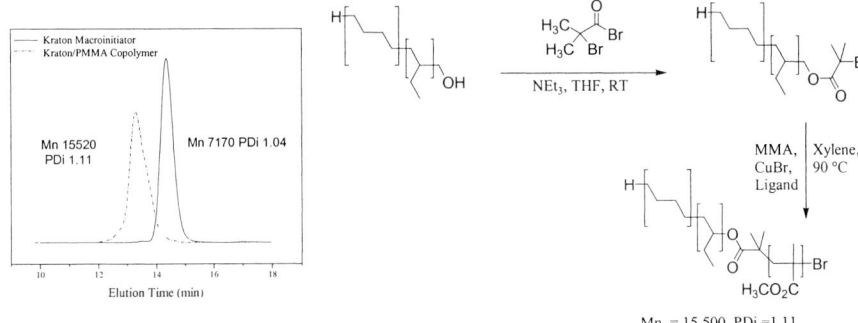

Fig. 3: Synthesis of octafunctional initiator and eight arm star polymers based on a sucrose initiator.

Block copolymers from macroinitiators

An important aspect of any living polymerisation technique is the ability to prepare block co-polymers. The transformation of alcohols into initiators can be exploited here, as

Fig. 4: Poly(ethylene/butylene-b-MMA) copolymer by polymerisation with macro-initiator form Kraton L-1201 under standard polymerisation conditions.

previously been demonstrated by Kops [11] and Matyjaszewski [12]. An example of this is the synthesis of a macro-initiator from α-hydroxyl functional poly(ethylene-butylene), a commercial product available from Shell (Kraton L-1201) [13], Fig. 4. This approach avoids complications arising from termination of the A block, found in all cases where sequential addition of monomers is employed. An added advantage of this approach is that initiator efficiency is always very high as termination by radical-radical reactions is not as prevalent with such high mass initiators as is found for small molecule initiators.

ω-Functional polymers from end-capping reactions

The terminal group of a polymer from atom transfer polymerisation is usually a tertiary bromide. This has been previously been transformed into more useful functionality via an azide group. An alternative approach is to use an end-capping reagent in the polymerisation that essentially uses a non polymerisable monomer, previously described by Sawamoto. We have been using this chemistry to end-cap a propagating polymer with protected functional trimethyl silyl enol ether, which is subsequently transformed into a hydroxyl on deprotection, Fig. 5. The hydroxyl group can be subsequently transformed into a macromonomer by condensation with, for example, methacryloyl chloride.

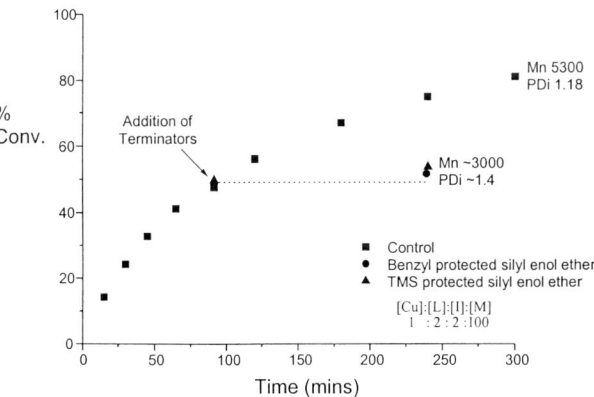

Fig. 5:End capping of PMMA with silyl enol ether to yield hydroxyl functional polymer

General Discussion

Transition metal mediated living polymerisation is an extremely versatile method to produce a wide range of polymers under relatively undemanding conditions. The use of protecting group chemistry can usually be avoided. This presentation has given an overview of some of the chemistry being developed to produce both α and ω functional polymers, star polymers, functional polymers and block co-polymers using copper(I) Schiff base catalysed chemistry. Full details of all of this chemistry will be the subject of future publications.

We wish to thank the EPSRC, The University of Warwick, Avecia, ICI, Unilever, BP and Biocompatibles for funding various aspects of this work.

References

1) M. Kato, M. Kamigaito, M. Sawamoto, T. Higashimura, *Macromolecules*, **28**(5), 1721 (1995).

2) M. Sawamoto, M. Kamigaito, *Trends Polym. Sci.*, **4**, 371 (1996).

3) J. S. Wang, K. Matyjaszewski, *JACS*, **117**(20), 5614 (1995).

4) D. M. Haddleton, A.J. Shooter, A. M. Heming, M. C. Crossman, D. J. Duncalf, S.R. Morsley, *ACS Symp. Ser.*, **685**, 284 (1998).

5) D. M. Haddleton, A.M. Heming, D. Kukulj, D. J. Duncalf, A.J. Shooter, *Macromolecules*, **31**, 2016 (1998).

6) K. Ohno, Y. Tsujii, T. Fukuda, *J. Pol. Sci. A-Pol. Chem.*, **36**(14), 2473 (1998).

7) D. M. Haddleton, M. C. Crossman, B. H. Dana, D. J. Duncalf, A. M. Heming, D. Kukulj, A.J. Shooter, *Macromolecules*, **32**, 2110 (1999).

8) D. M. Haddleton, D. Kukulj, D. J. Duncalf, A.H. Heming, A.J. Shooter, *Macromolecules*, **31**, 5201 (1998).

9) S. Angot, K. S. Murthy, D. Taton, Y. Gnanou, *Macromolecules*, **31**, 6748 (1998).

10) J. Ueda, M. Kamigaito, M. Sawamoto, *Macromolecules*, **31**(20), 6762 (1998).

11) K. Jankova, X. Y. Chen, J. Kops, W. Batsberg, *Macromolecules*, **31**(2), 538 (1998).

12) S. Coca, H.-J. Paik, K. Matyjaszewski, *Macromolecules*, **30**, 6513 (1997).

13) K. Jankova, J. Kops, X. Y. Chen, W. Batsberg, *Macromolecular Rapid Communications*, **20**(4), 219 (1999).

*Macromol. Symp. **157**, 209–216 (2000)*

Stereospecific Polymerization of α-Substituted Acrylates

Yoshio Okamoto*, Shigeki Habaue, Takahiro Uno, Hideo Baraki

Department of Applied Chemistry, Graduate School of Engineering,
Nagoya University, Nagoya 464-8603, Japan

SUMMARY: α-Substituted acrylates having various functional groups, such as alkoxymethyl, aminomethyl and alkylthiomethyl groups, on the α-position were synthesized and polymerized by radical and anionic methods, and stereoregularity of the obtained polymers was investigated. In the anionic polymerization using lithium reagents, highly isotactic polymers were obtained regardless of the polarity of solvents. Strong intra- and intermolecular coordination of the polar groups to counter-cation (Li^+) may be the main factor in controlling the stereochemistry. On the other hand, stereocontrol in the radical polymerization of α-(alkoxymethyl)acrylates was attained in the presence of zinc salts ($ZnBr_2$ and $ZnCl_2$), which may coordinate to the growing polymer and monomers.

Introduction

Design and stereospecific synthesis of polymers possessing a variety of functional groups are an attractive and important area in polymer science. α-Substituted acrylates are the monomers that have two different types of substituents on an olefin. Effects of polar groups on the reactivity and stereoregulation in anionic polymerization are of great interest from the view points of the intra- and intermolecular coordination to a counter-cation by the functional groups, and successful design of these groups can provide a useful method for the control of higher order structure, as well as the stereoregularity

$$H_2C = \underset{\underset{CO_2R}{|}}{\overset{\overset{CH_2FG}{|}}{C}}$$

FG: functional group

Results and Discussion

Stereospecific Anionic Polymerization of α-(Alkoxymethyl)acrylate

Although atactic polymers have been obtained in the radical polymerization of α-(alkoxymethyl)acrylates, little data are available about the tacticity of the anionic polymerization[1-3]. Thus, the anionic polymerization of benzyl α-(methoxymethyl)acrylate (**BMMA**) using lithium reagents was investigated[4]. The polymerization with alkyllithium at –78°C proceeded moderately, giving poly(**BMMA**) in 20 – 77% yields, while the radical polymerization with $(i\text{-PrOCO}_2)_2$ in toluene at 30°C afforded a polymer quantitatively.

In ^{1}H NMR spectra of the poly(**BMMA**)s obtained by the radical and anionic polymerizations in toluene (Fig. 1), the spectral pattern of the main chain methylene protons shows a typical AB quartet with a coupling constant 15.0 Hz for the latter polymer, indicating that the poly(**BMMA**) obtained by the anionic polymerization in toluene possesses a high isotacticity, while the polymer obtained by the radical method may be atactic. Fig. 2 shows the ^{13}C NMR spectra of the carbonyl carbon of poly(**BMMA**)s. A sharp peak is observed for the poly(**BMMA**) anionically prepared (Fig. 2(b)), and multi-peaks for the polymer obtained by the radical method (Fig. 2(a)). This observation also supports that the anionic polymerization of BMMA in toluene proceeds in a highly isotactic-specific manner. It is noteworthy that the polymers produced with a lithium reagent in THF showed a sharp peak assigned to highly isotactic sequences (Fig. 2(c)). This result is in marked contrast to that of the anionic polymerization of α-(alkyl)acrylates, which usually affords the polymers rich in syndiotacticity in THF[5-7].

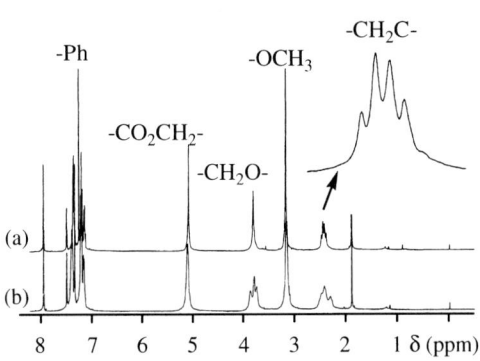

Fig. 1: ^{1}H NMR spectra of poly(**BMMA**)s obtained with t-BuLi in toluene at –78°C (a) and with a radical initiator at 30°C (b) (nitrobenzene-d_5, 110°C).

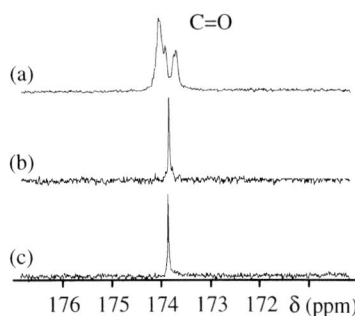

Fig. 2: ^{13}C NMR spectra of poly-(**BMMA**)s obtained with a radical initiator at 30°C (a), with t-BuLi in toluene (b) and with t-BuLi in THF (c) (CDCl$_3$, 60°C).

The anionic polymerization of α-substituted acrylates having various functional groups, such as alkoxymethyl, N,N-dialkylaminomethyl, 2-thienylmethyl and 2-pyridylmethyl groups, with lithium reagents proceeds in a highly isotactic-specific manner regardless of the polarity of solvents similarly to the anionic polymerization of **BMMA**[8 - 13]. Strong intra- and intermolecular coordination of the polar groups of the growing polymer chain and monomers to the counter-cation (Li$^+$), especially, a stable six-membered chelation of intermediate lithium enolate, should be the main factor in controlling the stereochemistry.

$$H_2C = \underset{\underset{CO_2R}{|}}{\overset{\overset{CH_2FG}{|}}{C}} \quad \xrightarrow[\text{toluene or THF}]{\text{lithium reagent}} \quad \left(\underset{\underset{CO_2R}{|}}{\overset{\overset{CH_2FG}{|}}{\underset{H_2}{C}} - \underset{}{C}} \right)_n$$

FG = OR', NR'R", SR',
2-thienyl, 2-pyridyl, etc.

highly isotactic

Stereospecific Polymerization of Chiral α-(Alkoxymethyl)acrylate

In the anionic polymerization of acrylate derivatives, introduction of a functional group on the α-position may be a novel method for the control of the higher order structure, in addition to the stereoregularity. A regular arrangement of successfully designed optically active groups along an isotactic polymer chain should form a chiral helical conformation[14, 15].

α-(Alkoxymethyl)acrylates bearing a chiral substituent, racemic and optically active benzyl α-(menthoxymethyl)acrylates ((±)- and (–)-**BMnMA**), were synthesized and polymerized by radical and anionic methods. The radical polymerization of **BMnMA** gave atactic polymers regardless of the chirality of the monomer, while the anionic polymerization of optically pure (–)-**BMnMA** and probably (±)-**BMnMA** provided highly isotactic polymers. The polymers prepared from optically pure (–)-monomer with an anionic initiator (the complexe of N,N'-diphenylethylenediamine monolithium amide (DPEDALi) with N,N,N',N'-tetramethylethylenediamine (TMEDA)) had low solubility in common organic solvents. The THF-soluble parts exhibited much lower specific rotation ($[\alpha]_{365}^{25} = -92°$: degree of polymerization (DP) = 16, Mw / Mn = 1.4) than those of the monomer ($[\alpha]_{365}^{25} = -226°$) and the radically obtained polymer ($[\alpha]_{365}^{25} = -201°$: DP = 14, Mw / Mn = 1.5).

The circular dichroism (CD) spectra of (–)-**BMnMA** and poly((–)-**BMnMA**)s obtained by radical and anionic methods are demonstrated in Fig. 3. Both CD spectra of the monomer (Fig. 3(a)) and radically obtained polymer (Fig. 3(b)) showed the negative cotton effect, which may be due to an isolated (–)-menthyl ether group, indicating that the optical activity of the radically obtained polymer is mainly based on that of each menthyl ether group on the side chain. The CD pattern of the anionically obtained polymer (Fig. 3(c)) was quite different from those of the monomer and the radically prepared polymer. In addition to a large positive cotton effect at 210 nm and a small one at 230 nm, peaks around 260 nm, which may be due to phenyl group were clearly observed. This indicates that a new chirality due to a regular arrangement of ether groups, probably a helical conformation of a polymer chain, is induced for the anionically obtained highly isotactic polymer. The lower specific rotation of the anionically prepared polymer arises from the compensation between the competing cotton effects of menthyl ether group and the helical structure.

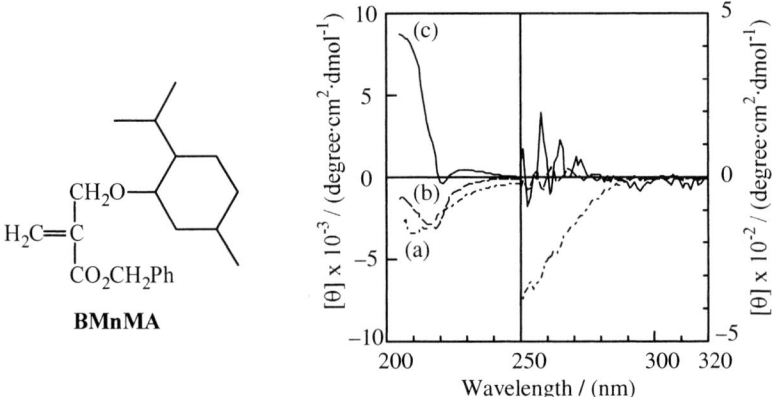

Fig. 3: CD spectra of the monomer, (–)-**BMnMA** (a), poly((–)-**BMnMA**) obtained with $(i\text{-PrOCO}_2)_2$ (b) and with DPEDA-TMEDA (c) (THF, r.t.).

The results of the anionic copolymerization of mixtures of (+)- and (–)-**BMnMA** initiated with DPEDALi-TMEDA complex are shown in Table 1. The copolymerization proceeded in a good yield in every case, and the solubility of the obtained copolymers decreased with an increase of enantiomeric excess (e.e.) of the charged monomer. Interestingly, non-linear relationship between the specific rotation of the copolymers and % e.e. of the charged monomer is clearly observed and the specific rotations reached a plateau value around 60 % e.e.. These results again suggest that the isotactic (–)-**BMnMA** sequence has a conformation that can contribute to the optical activity of the polymer chain.

Table 1. Anionic copolymerization of (+)- and (−)-
BMnMA with DPEDALi-TMEDA complex in toluene[a]

% e.e. of (−)-**BMnMA**	Yield (%)[b]	THF-soluble part			
		Yield (%)	DP [c,d]	Mw/Mn[d]	$[\alpha]_{365}^{25}$ [e]
0	89	100	20	1.32	—
20	92	100	14	1.49	−46°
40	90	100	15	1.74	−82°
60	79	100	15	1.85	−89°
80	95	80	16	1.55	−91°
100	94	16	16	1.38	−92°

[a] [monomer]/[initiator] = 20, temp. −78°C, time 24h.

[b] Methanol-insoluble part. [c] Degree of polymerization.

[d] Determined by SEC (polystyrene standard).

[e] In THF ($c = 0.5$).

Stereocontrol in Radical Polymerization of α-(Alkoxymethyl)acrylates

Stereocontrol in the radical polymerization of vinyl monomers is a most important goal in synthetic polymer chemistry. Although numerous works have reported on stereospecific polymerization based on anionic and coordination mechanisms, little is known concerning the stereocontrol in radical method[16 - 19]. Stereocontrol in radical polymerization using BMMA as a monomer in the presence of various metal salts, which may coordinate growing polymer and monomers, was investigated[20].

$$H_2C=\underset{CO_2Bn}{\overset{CH_2OMe}{C}} \xrightarrow[\substack{30°C,\ 48h}]{\substack{(i\text{-PrOCO}_2)_2 \\ -\ MX_n\ (1\sim1.5\ equiv.)}} \left(\underset{CO_2Bn}{\overset{CH_2OMe}{\underset{|}{C}}}\underset{}{\overset{}{\underset{H_2}{C}}}\right)_n$$

BMMA   poly(**BMMA**)

Fig. 4(b) shows the ^{1}H NMR spectrum of the main chain methylene protons of poly(**BMMA**) obtained by radical polymerization in the presence of ZnBr$_2$ in CH$_2$Cl$_2$, together with those of the polymers prepared without ZnBr$_2$ in CH$_2$Cl$_2$ (c) and with n-BuLi in toluene at −78°C ($m >$ 99%)[4] (c) for comparison. The spectral pattern of the polymer obtained in the presence of ZnBr$_2$ is quite different from that prepared without a salt, indicating that some stereocontrol takes place in the radical polymerization with ZnBr$_2$. While little change was observed on the stereoregularity of the polymers obtained with LiCl and MgCl$_2$, the polymerization using zinc salts (ZnCl$_2$ and ZnBr$_2$) gave the polymers having different tacticity from that prepared in the absence of metal salt. However, a polar solvent, THF, greatly reduced the effect of the stereocontrol by ZnBr$_2$ judging from the spectral pattern of ^{1}H NMR, suggesting that the coordination of zinc salts to the growing polymer and the monomer is important in this system.

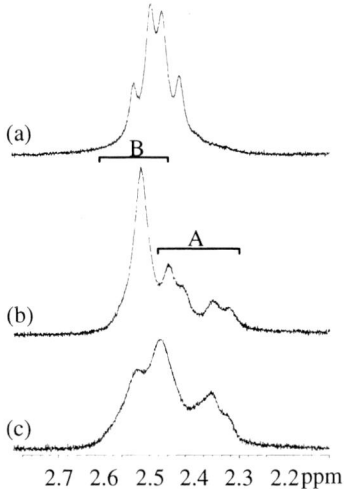

(a)

B

A

(b)

(c)

2.7 2.6 2.5 2.4 2.3 2.2ppm

Fig. 4: ^{1}H NMR spectra of the main chain methylene protons in poly-(**BMMA**) obtained with n-BuLi in toluene at −78°C (a), with radical method in the presence of ZnBr$_2$ in CH$_2$Cl$_2$ (b) (nitrobenzene-d_5, 110°C) and without ZnBr$_2$ (c) (nitrobenzene-d_5, 150°C).

In Fig. 4(b), the four peaks in the area A seem assignable to an AB quartet with a coupling constant of about 16 Hz due to the diastereotopic *meso*-protons. Therefore, a peak in the region B may be due to the protons of *racemo*-diad. However, since the chemical shifts for the former four peaks are different from those for the protons of the *isotactic*-poly(BMMA), the higher order (tetrad) tacticity should be considered the spectral pattern for the main chain methylene protons of poly(**BMMA**). Supposing that the peaks from 2.3 to 2.4 ppm are ascribed to one of the protons of *meso*-diad, diad tacticity of the polymer prepared using

ZnBr$_2$ is estimated to be $r : m = 67 : 33$, while that of the radically obtained polymer in its absence is nearly r ≈ m (≈ 0.5). The diad tacticities of the poly(**BMMA**) prepared in the presence of ZnBr$_2$ in toluene was evaluated to be $r : m = 70 : 30$.

The addition of zinc chloride accelerates the polymerization rate of methacrylates,[21] and tacticity is hardly changed during the polymerization, while an increase of isotacticity was observed in radical bulk polymerization of the complexes of methacrylate with zinc chloride.[22-24] The radical polymerization of benzyl methacrylate (**BnMA**) with ZnBr$_2$ was also carried out by the same procedure as that for the polymerization of **BMMA**. The polymer was obtained using ZnBr$_2$ (1.5 equiv.) in CH$_2$Cl$_2$ in 97% yield (DP = 59, Mw / Mn = 2.0) with a triad tacticity, $mm : mr : rr = 3 : 34 : 63$, while poly(**BnMA**) (96% yield, DP = 120, Mw / Mn = 4.8) prepared in its absence showed a tacticity, 3 : 33 : 64. These results indicate that the zinc salts characteristically affect the stereoregularity of poly(**BMMA**) in the radical polymerization, and a polar α-substituent on an acrylate is important in the stereocontrol of this system.

Conclusion

Stereospecific anionic and radical polymerizations of α-substituted acrylates having various functional groups, such as alkoxymethyl, aminomethyl and alkylthiomethyl groups, on the α-position were attained. In the anionic polymerization using lithium reagents, highly isotactic polymers were obtained regardless of the polarity of solvents. Strong intra- and intermolecular coordination of the polar groups to counter-cation (Li$^+$) for anionic polymerization or zinc salts for radical method should be the main factor in controlling the stereochemistry. New functional polymers with a variety of functional groups arranged on the main chain of a stereoregular polyacrylate were synthesized.

References
1. R. W. Lenz, K. Saunders, T. Balakrishnan, K. Hatada, *Macromolecules* **12**, 392 (1979)
2. T. Balakrishnan, R. Devarajan, M. Santappa, *J. Polym. Sci., Polym. Chem. Ed.* **22**, 1909 (1984)
3. G. Wulff, Y. Wu, *Makromol. Chem.* **191**, 3005 (1990)

4. S. Habaue, H. Yamada, Y. Okamoto, *Macromolecules* **29**, 3326 (1996)
5. H. Yuki, K. Hatada, *Adv. Polym. Sci.* **31**, 1 (1979)
6. H. Yuki, K. Hatada, T. Niinomi, K. Miyaji, *Polym. J.* **1**, 130 (1970)
7. K. Hatada, S. Kokan, T. Niinomi, K. Miyaji, H. Yuki, *J. Polym. Sci., Polym. Chem. Ed.* **13**, 2117 (1975)
8. S. Habaue, H. Yamada, T. Uno, Y. Okamoto, *J. Polym. Sci., Part A: Polym. Chem.* **35**, 721 (1997)
9. S. Habaue, T. Uno, Y. Okamoto, *Macromolecules* **30**, 3125 (1997)
10. S. Habaue, H. Baraki, Y. Okamoto, *Polym. J.* **29**, 872 (1997)
11. S. Habaue, T. Uno, H. Baraki, Y. Okamoto, *Polym. J.* **29**, 983 (1997)
12. S. Habaue, T. Shibagaki, Y. Okamoto, *Polym. J.,* in press.
13. H. Baraki, S. Habaue, Y. Okamoto, *Polym. J.* in press.
14. T. Uno, S. Habaue, Y. Okamoto, *Chirality* **10**, 711 (1998)
15. T. Uno, S. Habaue, Y. Okamoto, *Enantiomer*, in press.
16. For reviews,   see: K. Hatada, T. Kitayama, K. Ute, *Progress in Polymer Science* **13**, 189 (1988)
17. N. A. Porter, T. R. Allen, R. A. Breyer, *J. Am. Chem. Soc.* **114**, 7676 (1992)
18. T. Nakano, M. Mori, Y. Okamoto, *Macromolecules* **26**, 867 (1993)
19. K. Yamada, T. Nakano, Y. Okamoto, *Macromolecules* **31**, 7598 (1998)
20. S. Habaue, T. Uno, Y. Okamoto, *Polym. J.,* in press.
21. For example,   see: E. L. Mudruga, J. S. Roman, M. J. Rodriguez, *J. Polym. Sci., Polym. Chem. Ed.* **21**, 2749 (1983)
22. H. Hirai, T. Ikegami, S. Makishima, *J. Polym. Sci., Part A-1* **7**, 2059 (1969)
23. S. Okuzawa, H. Hirai, S. Makishima, *J. Polym. Sci., Part A-1* **7**, 1039 (1969)
24. T. Otsu, B. Yamada, M. Imoto, *J. Macromol. Chem.* **1**, 61 (1966)

Macromol. Symp. 157, 217–224 (2000)

Star-Shaped Nanomicelles
of Polyisobutylene-Polystyrene Diblock Copolymers.
New Stabilizer for Living Dispersion Polymerization of Styrene

Sándor Kéki[1], György Deák[1], Lajos Daróczi[2], Ákos Kuki[1], Miklós Zsuga[1]*

[1]*Department of Applied Chemistry,* [2]*Department of Solid State Physics*
Kossuth Lajos University, H-4010 Debrecen 10, Hungary

Summary

Formation of star-shaped nanomicelles of polyisobutylene-polystyrene (PIB-PS) diblock copolymers in hexane solution is reported. The length of the polystyrene segments were varied in the M_n range of 4000-13000 g/mol at approximately constant polyisobutylene segments length. The size and the size distribution of the nanomicelles were investigated by dynamic light scattering. Based on static light scattering measurements the mass-average molecular mass of the micelles and the number of arms were also determined. The synthesized diblock-copolymers were demonstrated to be capable of stabilizing the growing particles which were formed in the living anionic dispersion polymerization of styrene in hexane.

Introduction

When a linear block copolymer is dispersed in a block selective solvent for one of the blocks, micelles of well-defined morphology are formed. Micellization in dilute solutions of different block copolymers in selective solvents has attained great attention in the last years[1 11].

Micellization of block copolymers due to different nature of the blocks makes such copolymers very useful for the application as colloid stabilizers, emulsifiers and antifoaming agents. The same property makes such copolymers important in the emulsion, suspension and ionic dispersion polymerizations[13-15].

The size, the size distribution and the number of arms of the micelles formed in selective solvents determine the applicability of such copolymers.

In this paper we report a star-shaped micelle formation from the polyisobutylene (PIB)-polystyrene (PS) diblock copolymer in hexane, which is a good solvent for the PIB segments and bad for the PS segments. Our aim was to study the size, the size distribution, the number of arms and the shape of micelles formed form the PIB-PS diblock copolymer as a function of the length of the PS segments. The resulted PIB-PS diblock copolymers were tested as steric stabilizers in the living anionic dispersion polymerization of styrene.

Experimental
Synthesis of polyisobutylene-polystyrene diblock copolymers
The block-copolymerizations were performed in a 1000 ml three-necked flask in a Dry-Box, at -80 °C, under dry nitrogen atmosphere. The reaction mixture was stirred with a mechanical

        CCC 1022-1360/00/$ 17.50+.50/0

stirrer. Isobutylene, pyridine and 2-chloro-2,4,4-trimethylpentane (TMPCl) were dissolved in a mixture of methylcyclohexane/CH_2Cl_2 (60/40 v/v). The volume of the solution was 600 ml. The polymerization was started by adding $TiCl_4$ as a coinitiator to the solution. After a period of time, "in situ" styrene was added to the reaction mixture. The polymerization was terminated with precooled methanol. The polymer was purified by successive precipitation from its CH_2Cl_2 solution with methanol.

Anionic dispersion polymerization of styrene in hexane

After dissolution of the diblock-copolymers (0.2 g) styrene (5 ml) was added to the solution. The polymerizations were initiated by adding 0.6 ml of 2.5 mol/l n-buthyllithium (nBuLi) to the reaction mixture. After a predetermined interval, the polymerization was terminated with 2 ml of methanol. The dispersions were investigated by dynamic light scattering (DLS), light microscopy and by transmission electronmicroscopy (TEM) in order to determine the size of the particles formed.

Instruments

A Waters Size-Exclusion Chromatograph (SEC) equipped with Waters differential refractometer, UV-detector and five Ultrastyrogel columns (7.8 x 300 mm) was applied for the determination of the molecular weights and molecular weight distributions. The system was calibrated with polystyrene and polyisobutylene standards. Tetrahydrofuran was used as the eluent.

The light scattering experiments were performed with a Brookhaven Light Scattering device equipped with a BI-9000 digital correlator. The light source is a solid state, vertically polarized laser operating at 533.4 nm. All measurements were performed at 25 °C.

The refractive indices (dn/dc) of the hexane solutions of the diblock were measured with a Rayleigh- type interferometer (Carl Zeiss, Jena, Germany) at 538 nm.

The light microscope experiments were made on Zeiss AxioTech type microscope. The TEM pictures were taken by a Jeol 2000 FX-II type transmission electron microscope.

Results and Discussion

Formation of nanomicelles from polyisobutylene-polystyrene diblock copolymers in hexane

The dynamic light scattering experiments were performed in tetrahydrofuran (which is a good solvent for both segments) and hexane (which is a bad solvent for the PS segment but good for the PIB segment). The correlation functions on solutions of the PIB-PS diblock were obtained at least in four different angles in the range of 60-155° and were analyzed by the method of cumulants and the NNLS method (Non-negative Constraint Least Squares). The determined effective diameters by DLS in tetrahydrofuran were in the range of 4-5 nm. However, when dissolving the copolymers in hexane, much larger sizes were obtained indicating the aggregation of the PIB-PS copolymers. A typical micelle size distribution for a PIB-PS block copolymer measured in hexane at 90° scattering angle is presented in Fig. 1. In all cases the size distributions were unimodal in the concentration range applied (0.5-2 mg/ml).

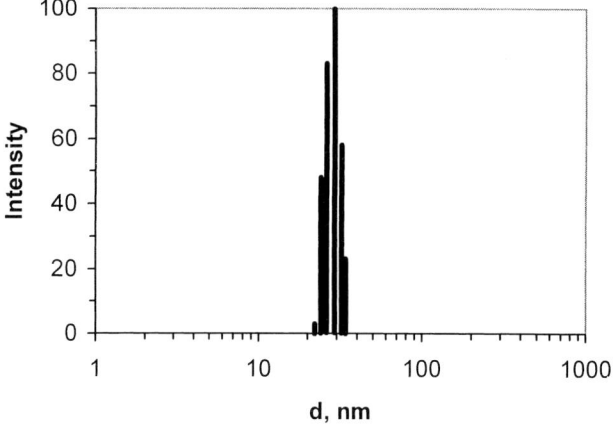

Fig. 1. Size distribution determined by DLS of micelles formed from the PIB-PS diblock in hexane, ($\Theta = 90°$). The length of the PS and the PIB segments are 4300 and 11000 g/mole by SEC and ^{1}H-NMR, respectively. **The effective diameter: 29 nm. The polydispersity: 0.11.**

Estimation of the aggregation number and number of arms

Plotting the diffusion coefficients and the effective diameters versus the concentration of the PIB-PS diblock copolymer in the range of 0.5-2 mg/ml no significant dependence can be recognized (Fig. 2).

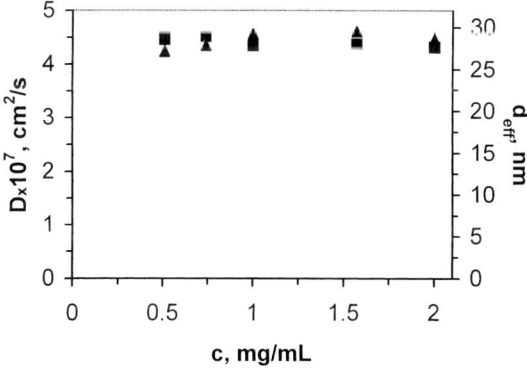

▲ Diffusion coefficient ■ Effective diameter

Fig. 2. Dependence of the effective diameter and diffusion coefficient on the concentration of the PIB-PS diblock in hexane. The length of the PS and the PIB segments are 4300 and 11000 g/mole by SEC and ^{1}H-NMR, respectively.

This indicates that the size of the micelles in the concentration range applied does not vary significantly with the concentration of the PIB-PS block copolymer. This observation, i.e., that the micelle sizes is independent of the copolymer concentration, is characteristic of micelle formation by a closed association process. Once the maximum size of the micelle has been reached a further increase in the block copolymer concentration causes the increase of the concentration of the micelles with the same size. Therefore, the aggregation number of the micelle can be determined by static light scattering experiments by means of the Zimm-plot analysis.

The aggregation number and the number of arms can be expressed by Eq. 1.

$$f = M_{w,a}/M_{w,u} \tag{1}$$

where f is the number of unimers in the micelle, i.e., the aggregation number; $M_{w,a}$ and $M_{w,u}$ are the weight average molecular mass of the aggregate and the unimer, respectively.

$M_{w,a}$ were determined by static light scattering measurements using the Zimm-plot analysis. The number of arms (n_a) for the micelle in the case of the PIB-PS diblock copolymer is $n_a = f$. Theoretically, the R_H/R_g values should be 1.29 for a homogenous sphere (where R_H = hydrodynamic radius and R_g = radius of gyration). The R_H/R_g values are close to this value (see Table 1.) which support the formation of spherical micelles.

Estimation of the size of the hard core of the micelles

The size of the hard core of the micelles (R_c) can be estimated by the following formula:

$$[3M_{w,a}w_{PS}/(4\pi N_A\rho_{PS})]^{1/3} = R_c \tag{2}$$

where $M_{w,a}$ is the mass-average molecular mass of the micelles; w_{PS} is the mass fraction of polystryrene in the diblock; ρ_{PS} is the density of dry polystyrene (12) (which is 1.05 g/cm³); and N_A is the Avogadro's number.

The dependence of the size of the hard core (d_c), the hydrodynamic diameter (d_h) and the number of arms of the micelles (f) on the molecular mass of the polystyrene segments (M_n) are presented in Table 1. and Fig. 3. The size of the hard core is levelling off on increasing molecular mass of the PS block. This limits the aggregation number, i.e., the number of arms. The lower the frequency of the arms is the higher the possibility of solvation is, i.e., the hydrodynamic diameter is increasing with the molecular mass of the hard core segments.

Living anionic dispersion polymerization of styrene in hexane in the presence of the PIB-PS diblock copolymer. A new dispersing agent

Our goal was to study the effect of the PIB-PS diblock as stabilizer on the dispersion polymerization of styrene. Anionic dispersion polymerization of styrene in hexane was carried out in the presence of PIB-PS diblock copolymers of varying PS segments in length.

Code	$M_{n[PS]}$ (g/mol)	$M_{n[PIB]}$ (g/mol)	d_c (nm)	d_h (nm)	f	R_h/R_g
PIBPS1	4300	11000	7.8	29	29	1.35
PIBPS2	6500	10300	10.4	33	46	1.39
PIBPS3	9040	8900	10.6	42	34	1.22
PIBPS4	13600	11400	10.7	45	27	1.4

Table 1. The diameter of the hard core, the hydrodynamic diameter and the number of arms of the micelles in hexane solution.

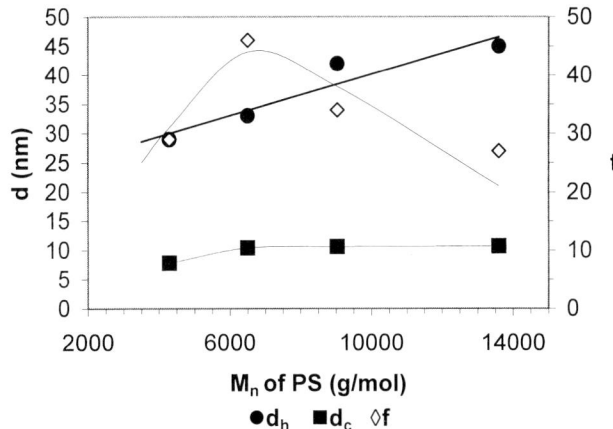

Fig. 3. The dependence of the size of the hard core, the hydrodynamic diameter and the number of arms of the micelles on the molecular mass of the polystyrene (hard) segments in hexane solution.

Without diblock copolymers, a very rough PS precipitate was obtained. By increasing the length of the polystyrene in the diblock, narrower and smaller size of polystyrene particles were obtained under the same conditions. The images obtained with optical microscope and TEM are shown in Fig. 4. This tendency is probably due to the greater adsorption energy of the large polystyrene segments onto the surface of the growing polystyrene particles. In the case of the shorter polystyrene segment the adsorption energy is not sufficient, therefore desorption of the diblock can take place resulting in large particle size and broad particle size distribution. However, the M_n and M_w/M_n of the resulted polystyrene produced with different PIB-PS diblock stabilizers did not change significantly. The M_n of the polystyrene increases with the monomer conversion. Typical SEC traces are shown in Fig.5. The dispersion polymerization seems to be slowly initiated living polymerization. The livingness of the polymerization is proved by plotting the diagnostic lines, i.e., the M_n versus W_p and $-\ln(1-C)$ versus time lines (Fig. 6). To interpret the effect of the stabilizer, we postulated that the size of the hard core of the micelles formed from the PIB-PS diblock copolymers is increased by styrene. Therefore, independent DLS experiments were performed on the solution of the diblock copolymers in the presence of styrene. However, these experiments did not show an increase of the size of the micelles, i.e., no swelling of the micelles takes place.

Probably, the nanomicelles originally present in hexane, are collapsed in the presence of polystyrene nuclei, i.e., the unimers of the PIB-PS diblock are attached onto the polystyrene surface. Thus, there should exist an equilibrium between the nanomicelles and unimers (shifted to micelles in the absence of polystyrene particles) and unimers adsorbed onto the polystyrene particles (Scheme 1).

After evaporation the solvent, the bulk polystyrene samples obtained can be re-dispersed to the extent of their initial size, i.e., as they were before preparation, indicating that the stabilizer molecules are attached firmly onto the surface of the polystyrene particles.

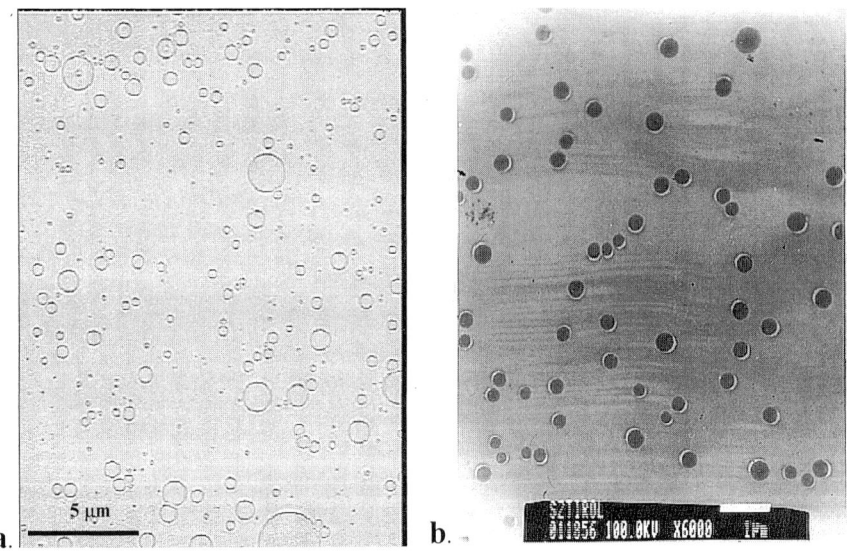

a. b.

Fig. 4. Optical microscope (a) and transmission electronmicroscope micrograph (b) of polystyrene particles prepared via anionic dispersion polymerization.
Conditions: 4 ml styrene, 0.002 mol nBuLi and 0.2 g **PIBPS1** (a) and 0.2 g **PIBPS4** (b) in 45 ml hexane. T=30 °C, reaction time= 60 min. Composition for **PIBPS1** and **PIBPS4** are in Table 1.

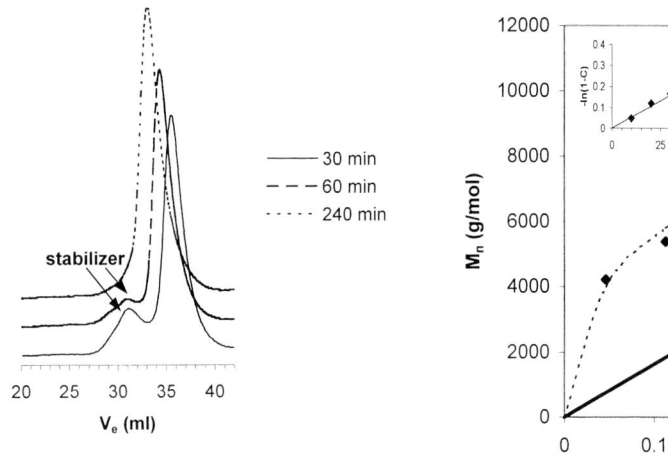

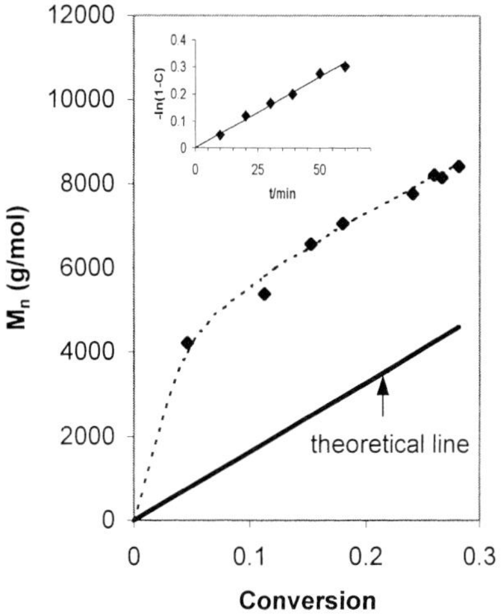

Fig. 5. SEC traces of polystyrene prepared by anionic dispersion polymerization
Conditions: 4 ml styrene, 0.002 mol nBuLi and 0.2 g **PIBPS4** in 45 ml hexane. T=30 °C

Fig. 6. Diagnostic plots for proving the livingness of anionic dispersion polymerization of styrene.
Conditions: 10 ml styrene, 0.005 mol nBuLi and 0.5 g **PIBPS4** in 115 ml hexane. T=30 °C

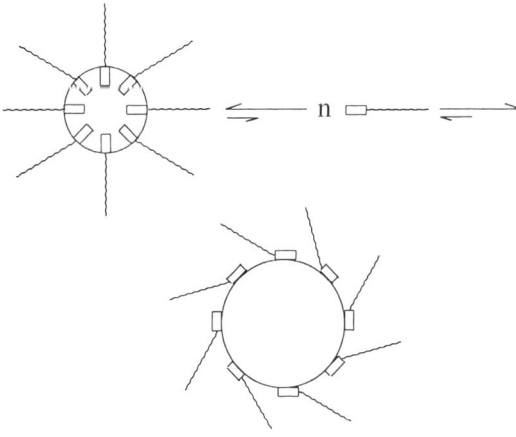

Scheme 1. Schematic representation of stabilization of polystyrene particles by the PIB-PS diblock copolymers

Conclusion

Formation of star-shaped nanomicelles were recognized in hexane solution of the PIB-PS diblock copolymers. The size and size distribution of the particles were determined by dynamic light scattering (DLS) measurements. On the basis of dynamic and static light scattering, spherical, compact nanomicelles formation is concluded.

The synthesized PIB-PS diblock copolymers were tested as steric stabilizers in the living dispersion anionic polymerization of styrene, and the **PIBPS4** copolymer was proved to be the best stabilizer.

Acknowledgment

This work was financially supported by the grants Nos. T 019508, T 025379, T 025269, T 030519, M 28369 given by OTKA (National Found for Scientific Research Development, Hungary) and by the grants of FKFP 04441/1997 and OMFB MEC 00699/99.

References

(1) Kéki, S.; Bogács, L.; Bogács, Cs.; Daróczi, L.; Zsuga, M. *Die Angewandte Makromolekulare Chemie*, **245**, 183 (1997)
(2) Zhou, Z.; Chou, B.; Peiffer, D.G. *Macromolecules*, **26**, 1876 (1993)
(3) Astafieva, I.; Khougaz, K.; Eisenberg, A. *Macromolecules*, **28**, 7127 (1995)
(4) Prochazka, K.; Martin, T.; Munk. P.; Webber, S.E. *Macromolecules*, **29**, 6518 (1996)
(5) Cogen, K.A.; Gast, A.P.; Capel, M. *Macromolecules*, **24**, 6512 (1991)
(6) Smith, C.K.; Liu, G. *Macromolecules*, **29**, 2060 (1996)
(7) Balsara, N.P.; Tirrell, M.; Lodge, T.P. *Macromolecules*, **24**, 1975 (1991)
(8) Raspaud, E.; Lairez, D.; Adam, M.; Carton, J.P. *Macromolecules*, **27**, 2956 (1994)
(9) Tao, J.; Stewart, S.; Liu, G.; Yang, M. *Macromolecules*, **30**, 2738, (1997)
(10) Hurtrez, G.; Dumas, P.; Reiss, G. *Polymer Bull.*, **40**, 203 (1998)
(11) Kéki, S.; Deák, Gy.; Kuki, Á; Zsuga, M. *Polymer,* 39, 6053 (1998)
(12) Brandrup, J.; Immergut, H.E.: Polymer Handbook, p. V. 59, Wiley-Interscience Publication, New York-London-Sidney-Toronto (1975)
(13) Imperial Chemical Industries, British Pat. 893, 429 (1962)
(14) Rohm & Hass, British Pat. 934, 038 (1963)
(15) Awan, M.A.; Dimonie, V.L.; El-Aasser, M.S. *J. Polym. Sci. Part A.*, **34**, 2633 (1996)

Synthesis and Solution Properties of Star-Shaped Poly(*tert*-butyl acrylate)

Daniela Held[a] and Axel H. E. Müller[a,b]*

[a] Institut für Physikalische Chemie, Universität Mainz, Welderweg 15, D-55099 Mainz, Germany

[b] Makromolekulare Chemie II and Bayreuther Institut für Makromolekülforschung, Universität Bayreuth D-95440 Bayreuth, Germany. E-mail: axel.mueller@uni-bayreuth.de

Abstract:

A series of star polymers consisting of poly(*tert*-butyl acrylate) arms and an ethylenegly-col dimethacrylate (EGDMA) microgel core were synthesized using anionic polymeriza-tion. The effect of various parameters (precursor length, ratio [EGDMA]/[Initiator], reac-tion time, and overall concentrations) on the average number of arms was investigated. Molecular weights were determined using GPC coupled with an online viscometer and MALLS. The exponents for the relation between intrinsic viscosity or radius of gyration and molecular weight, respectively, are extremely low, indicating that the dimensions of the star polymers only slightly increase with the number of arms. After a cer-tain number of arms is reached the intrinsic viscosity even decreases with molecular weight. Computer simulations for star polymers were carried out where the radius of gyration was calculated as a function of the number of arms. The results are in good agreement with the experimental data.

Introduction

Branched structures have been the subject of continuing interest in polymer chemistry. Besides randomly branched, comb-shaped and hyperbranched polymers, star-shaped poly-mers have been investigated for several years[1]. Star polymers consist of linear chains connected to one single junction point referred to as core. There are different ways to obtain a star polymer using controlled polymerization. In order to obtain a defined number of arms per molecule, multifunctional initiators or termination reagents have been used, mostly to form polystyrene or polydiene stars[2]. However, this method was not really successful for polar monomers like (meth)acrylates, because of the reduced reactivity towards the multifunctional termination reagents at the low temperatures needed for controlled polymerization[3]. Another possibility is the use of a difunctional monomer like divinylbenzene, either to form a pluri-functional initiator by reaction with a monofunctional one (core first)[4] or by coupling with active chain ends (arm first)[5]. Stars produced in such a way have a broad distribution of the number of arms. An advantage of the arm-first method is that it is possible to isolate and characterize the polymer chains which become the arms of the resulting star. The average number of arms, f, is therefore easily determined when the molecular weight of the star polymer is known.

During the last decade PMMA star polymers were synthesized using GTP or anionic polymerization[6,7]. The bifunctional monomer used here was ethyleneglycol dimethacrylate (EGDMA). The polymers were often characterized by GPC coupled with viscosity or multiangle laser light scattering (MALLS) detectors. Poly(*tert*-butyl acrylate) (P*t*BuA) star polymers were also prepared but here a different method was used[8]. Short polystyrene chains were coupled with divinylbenzene (arm-first) and the resulting active centers in the core were used to initiate the living anionic polymerization of *t*BuA. Therefore star polymers with two different kinds of arms and a very non-polar core were achieved. The effect of various parameters, i. e., precursor length, ratio [EGDMA]/[Initiator] ($[E]_0/[I]_0$) and overall concentration, on star formation and on the average number of arms, f, is still under discussion in literature. Mays[9], Burchard[10] and Higashimura[11] and their coworkers found that the number of arms increases with the ratio $[E]_0/[I]_0$. The two latter groups also claimed a decrease of f with increasing arm molecular weight. For star polymers produced by GTP Haddleton and Crossman[12] found that the number of arms is not related to the arm length. For the stars made by anionic polymerization they emphasize the important role of the overall concentrations[13]. Higher overall concentrations lead to a higher average number of arms.

Compared to their linear analogues with the same molecular weight branched polymers have smaller dimensions leading to a reduced viscosity. This effect becomes more pronounced with increasing number of branches. Nevertheless, the exponent of the relation between intrinsic viscosity and molecular weight, $[\eta] = K \cdot M^{\alpha}$, is the same for a linear polymer and a star with a constant number of arms[14]. This means that in a Mark-Houwink plot a star polymer which grows by adding monomer to the arms, results in a parallel line to the linear one, but shifted to lower viscosity. The Mark-Houwink exponent, α, depends on the structure of the polymer in solution. Its value can be in the range from 0 (solid sphere) to 2 (rigid rod). For the relation between the radius of gyration and molecular weight of star polymers in good solvent Daoud and Cotton[15] calculated a dependence of $R_g \propto N^{0.6} f^{0.2}$ (N = number of segments per arm). For N = const. this results in $R_g \propto M^{0.2}$. Stars made by the arm-first approach with a bifunctional monomer have an interesting feature, because they do not have a constant number of arms. Here the molecular weight increases by adding arms and therefore the structure changes within the sample. This should result in a smaller exponent in a plot analogous to a Mark-Houwink plot[16].

In this work star polymers with P*t*BuA arms were prepared via the arm-first strategy using EGDMA as a linking agent. P*t*BuA stars are convenient starting materials for the synthesis of poly(acrylic acid) stars. Furthermore, the molecular parameters in solution, the exponent of the intrinsic viscosity vs. molecular weight, α, and the exponent for the relation between the radius of gyration and molecular weight, α_s, were determined using a viscometer and MALLS coupled with GPC.

Experimental Part

Reagents: Diphenylethylene (DPE) was purified by adding n-butyllithium until the solution turned slightly red followed by distillation. Diphenylhexyllithium (DPHLi) was prepared by adding n-butyllithium (1.6 M solution in hexane) to an excess of DPE in hexane. The solution was stirred at room temperature until the DPHLi precipitated. The red solid was isolated and washed with hexane several times. At last the initiator was dried on the vacuum line and stored under nitrogen. LiCl (Merck) was dried in high vacuum at 300 °C for two days. *tert*-Butyl acrylate (*t*BuA, BASF AG) was fractionated from CaH_2 over a 1 m column filled with Sulzer packing at 45 mbar, stirred over CaH_2, degassed and distilled in high vacuum just before use. Ethyleneglycol dimethacrylate (Röhm GmbH) was fractionated over a 10 cm Vigreux column, stirred over CaH_2, distilled in high vacuum and filtrated over a short column filled with dry neutral alumina. Decane (internal standard for GC, Aldrich) was stirred over sodium/potassium alloy, degassed and distilled on the vacuum line. THF (BASF AG) was fractionated over a 1.5 m column, stirred twice over sodium/potassium alloy, degassed and distilled on the vacuum line.

Polymerization: All experiments were carried out in a stirred tank reactor under nitrogen atmosphere. The initiator ($[I]_0 = 1 \cdot 10^{-3}$ mol/l) and a fivefold molar excess of LiCl dissolved in THF were introduced and cooled to −78°C. Then *t*BuA was added. After 5 min reaction time a sample (precursor) was taken and EGDMA was introduced. After 10 min. a first sample was taken. The polymerization was quenched after 90 min with acidic methanol. Conversion of EGDMA (x_E) was measured by gas chromatography. The polymers were isolated by precipitation into methanol/water (50/50). It was dried under vacuum at room temperature for at least 2 days.

Characterization: GPC measurements were performed at room temperature using 5 μ PSS SDV gel columns (column set: 10^3, 10^5, 10^6 Å) with THF as eluent at a flow rate of 0.5 ml/min. Detectors: Applied Biosystems 1000S UV diode-array detector; Wyatt Technology DAWN-DSP F multiangle laser light scattering instrument equipped with an He-Ne laser ($\lambda = 633$ nm); Viscotek viscosity detector H 502B; Shodex refractive index detector (tungsten lamp) or NFT ScanRef scanning refractometer ($\lambda = 633$ nm) for online-determination of the refractive index increment. The average molecular weights of the star polymers were determined using GPC-Viscosity coupling.

Simulations: Monte-Carlo simulations were performed on a diamond lattice using a program developed by W. Radke[17].

Results and Discussion

Synthesis

A series of star polymers with PtBuA arms were synthesized using EGDMA as a linking agent. The precursor length was one of the investigated parameters. Therefore two different precursors with $M_w \approx 20,000$ and $M_w \approx 90,000$ were synthesized. The reaction time chosen for all precursors was 5 min. Efficient stirring is necessary because the reaction is extremely fast. The reaction time for the linking agent EGDMA was 90 minutes, after 10 minutes a sample was taken. All samples were characterized using GPC coupled with a viscometer and a MALLS detector. The refractive index increment dn/dc, was determined using a scanning interferometer. The samples were not fractionated which means that unreacted precursor is still present. However the average molecular weights of the star polymers could be determined by subtracting the residual precursor from the GPC eluograms. The number- and weight-average numbers of arms were calculated as

$$f_n = M_{n,star} \cdot (1-w_{core})/M_{n,arm} \quad \text{and} \quad f_w = M_{w,star} \cdot (1-w_{core})/M_{n,arm}.$$

The weight fraction of the core was calculated as $w_{core} = m_E \cdot x_E/[m_E x_E + m_{tBuA} \cdot (1-w_{prec})]$ where w_{prec} is the fraction of residual precursor in the eluogram.

First, the ratio $[E]_0/[I]_0$ was varied for two different arm molecular weights while the arm length, the reaction time and the initiator concentration were kept constant. For the low molecular weight precursors star polymers with only 3 - 6 % residual arms are formed after 90 min, as shown in Figure 1.

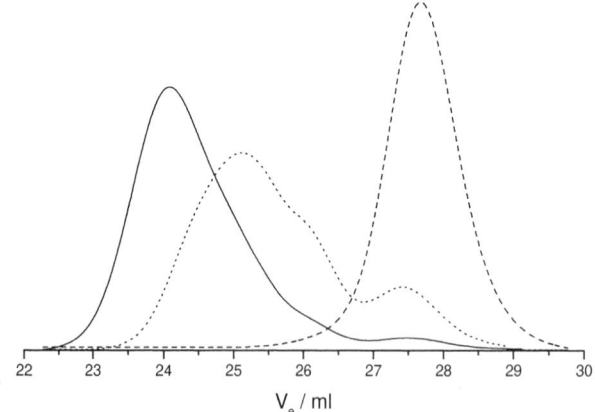

Figure 1: GPC eluograms of a low molecular weight precursor ($M_w = 20,000$) and the resulting star polymers after 10 and 90 min; $[E]_0/[I]_0 = 15$

For the high molecular weight precursors the content of residual arms decreases from 32% to 18% with increasing ratio $[E]_0/[I]_0$. Figure 2 shows the GPC eluograms of a sample with high precursor molecular weight and $[E]_0/[I]_0 = 15$ for different reaction times. The star formation is significantly slower; after 10 minutes most of the polymer consists only of precursor, dimer and trimer.

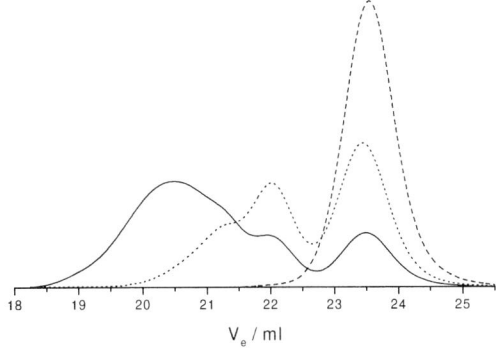

Figure 2: GPC eluograms of a high molecular weight precursor (---) and the resulting star polymers after 10 (·····) and 90 (——) min; $[E]_0/[I]_0 = 15$

Table 1 shows the dependence of the average number of arms on various parameters. f_w decreases with the length of the precursor and increases with reaction time and ratio $[E]_0/[I]_0$.

Table 1: Experimental conditions and results for star polymers with short and long precursor for different ratios $[EGDMA]_0/[I]_0$ and different reaction times. $[I]_0 = 1 \cdot 10^{-3}$ mol/l

$M_{n,Pre} \cdot 10^{-4}$ (M_w/M_n)	$[E]_0/[I]_0$	t, min	$M_n \cdot 10^{-4}$	$M_w \cdot 10^{-4}$	$1-w_{core}$	f_n	f_w	x_E	w_{Prec}
1.80	5	10	4.08	4.63	0.88	1.99	2.26	0.77	0.42
(1.08)		90	6.43	8.18	0.90	3.22	4.10	1.00	0.60
1.71	10	10	6.29	7.93	0.84	3.09	3.89	0.81	0.10
(1.09)		90	8.65	11.5	0.83	4.20	5.59	1.00	0.04
1.70	15	10	5.91	7.30	0.81	2.81	3.47	0.64	0.11
(1.09)		90	9.61	12.9	0.78	4.39	5.88	0.94	0.03
9.64	5	10	18.6	20.7	0.94	1.82	2.03	0.69	> 0.50
(1.06)		90	27.1	37.8	0.97	2.74	3.82	0.90	0.32
10.5	10	10	21.3	27.5	0.95	1.93	2.49	0.65	0.48
(1.03)		90	31.5	47.2	0.96	2.87	4.30	0.89	0.23
10.5	15	10	22.0	26.0	0.93	1.95	2.31	0.60	0.50
(1.04)		90	36.2	54.8	0.94	3.24	4.91	0.83	0.18

For the high molecular weight precursors the conversion of EGDMA was not complete after 90 minutes reaction time. Therefore experiments were performed where the reaction time was increased up to 24 hours. This leads to nearly full conversion of the crosslinker. The average number of arms still increased with time while the content of residual precursor decreased (Table 2). Furthermore, an experiment was carried out, where the overall concentrations of the reagents was doubled ($[I]_0 = 2 \cdot 10^{-3}$ mol/l). This had a strong effect on the average number of arms. f_w is nearly twice as high as compared to the experiment with the low concentrations.

Table 2: Experimental conditions and results for star polymers with long precursors and ratio $[E]_0/[I]_0 = 15$

$M_{n,Pre} \cdot 10^{-4}$ (M_w/M_n)	$[I]_0 \cdot 10^3$, mol/l	t, min	$M_n \cdot 10^{-4}$	$M_w \cdot 10^{-4}$	$1-w_{core}$	f_n	x_E	w_{Prec}
8.60 (1.04)	1	10	15.6	17.6	0.93	1.69	0.55	0.57
		90	27.8	39.0	0.94	3.05	0.86	0.17
		1275	44.9	63.6	0.94	4.91	0.97	0.11
7.94 (1.04)	2	10	20.4	26.9	0.95	2.45	0.62	0.28
		90	37.4	56.7	0.94	4.44	0.90	0.13
		1440	74.0	105	0.94	8.76	1.00	0.10

The eluograms for the experiments with the higher initiator concentration are shown in Figure 3. A shoulder is observed at extremely high molecular weight. This is assigned to a coupling reaction of two star molecules occurring between the anions of one core with unreacted vinyl groups of another one.

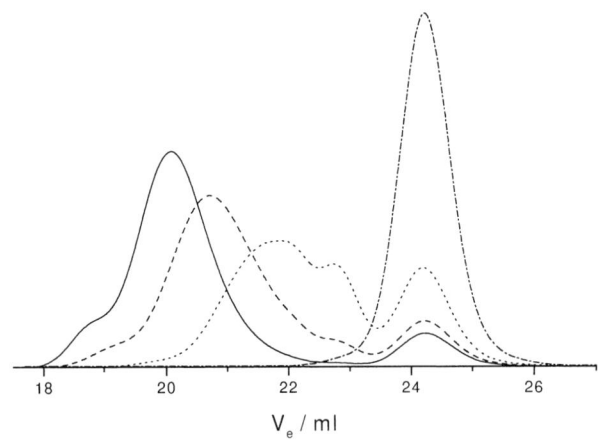

Figure 3: GPC eluograms; precursor (-·-·), 10 (·····), 90 (---) and 1440 (——) min reaction time with EGDMA; $M_{w,Pre} = 72000$; $[E]_0/[I]_0 = 15$; $[I]_0 = 2 \cdot 10^{-3}$ mol/l

Characterization

Intrinsic viscosity. Two different methods were used to obtain a Mark-Houwink plot and to calculate the exponent, α. First, it was determined from GPC coupled with a viscometer using a sample with a broad molecular weight distribution and, alternatively, from M_w and $[\eta]$ of various star polymers of the same run taken at different reaction times. In both cases extremely small values ($0 < \alpha < 0,1$) are obtained, even when the average number of arms is low. The results are summarized with the results from GPC-MALLS in Table 3. The plots (Figure 4 and Figure 5) (both 90 min reaction time) also show the signal of the specific viscosity and the contraction factor g' = ($[\eta]_{branched}/[\eta]_{linear})_M$. Independently of the arm molecular weight the intrinsic viscosity decreases for star polymers down to 20% compared to a linear polymer with the same molecular weight. Figure 5 shows data for a star obtained from a precursor with high molecular weight. The signals for the precursor, the dimer, and trimer are still present after 90 min reaction time for crosslinking.

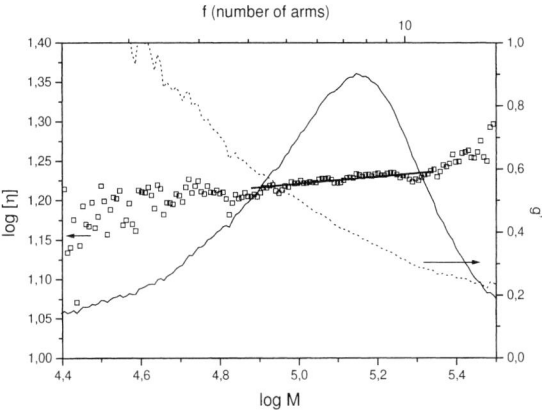

Figure 4: Mark-Houwink plot for a star polymer with short precursors ($M_w = 20,000$) and $[E]_0/[I]_0 = 15$; intrinsic viscosity (□), specific viscosity (—) and contraction factor, g' (·····), in dependence of molecular weight (or number of arms) in THF at 30 °C; $\alpha - 0.045$.

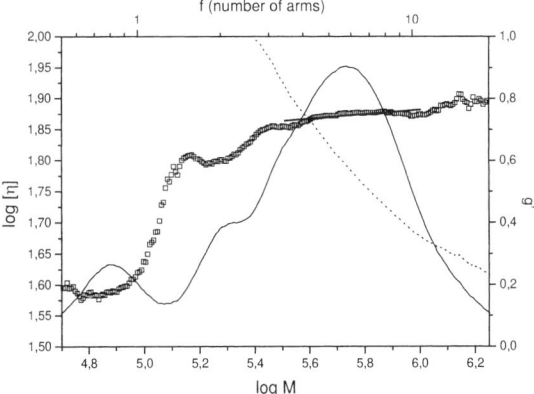

Figure 5: Mark-Houwink plot for a star polymer with long precursors ($M_w = 90,000$) and ratio $[E]_0/[I]_0 = 15$; $\alpha - 0.036$.

232

For star polymers with high molecular weight (high number of arms) we even observe negative exponents (Figure 6), which means that stars with higher molecular weight have a lower
intrinsic viscosity.

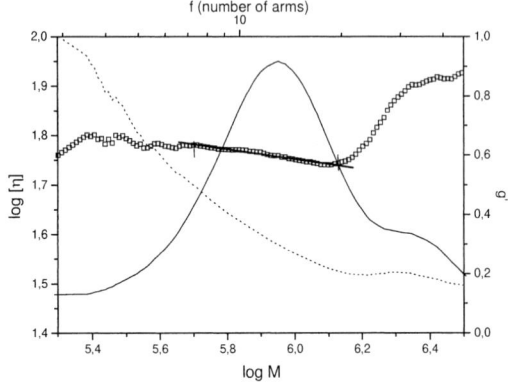

Figure 6: Mark-Houwink plot for a star polymer with long precursors (M_w = 90,000)
and $[E]_0/[I]_0$ = 15; $[I]_0$ = 2 x10-3 mol/l; α = -0.10.

It seems that for high molecular weight stars a negative exponent is obtained and for stars
with less arms (and low molecular weight) a small, positive exponent. Therefore there should
be a maximum in the plot of intrinsic viscosity vs. molecular weight. To detect the maximum
of the intrinsic viscosity a mixture of samples taken at different reaction times was used to
achieve a sample with a broad arm number distribution. In Figure 7 the maximum of the
intrinsic viscosity, which is also implied in Figure 5, can be clearly seen.

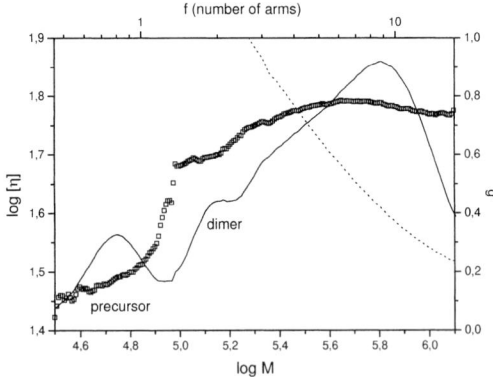

Figure 7: Mark-Houwink plot for a mixture of star polymer samples after 10, 90 and
1440 minutes reaction time; $[E]_0/[I]_0$ = 15; $[I]_0$ = 2 x10^{-3} mol/l.

The maximum of the intrinsic viscosity can also be seen in the plot of log [η] vs. log M_w for
samples of the same precursor taken at different reaction times (Figure 8).

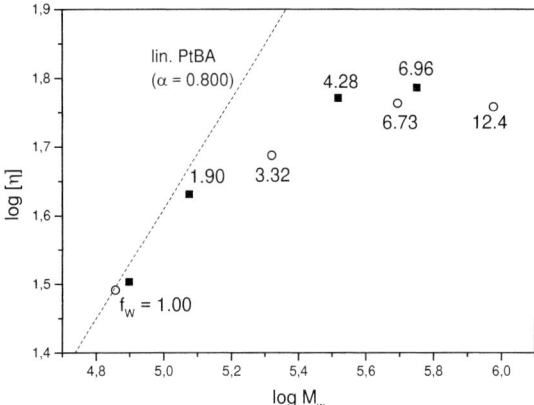

Figure 8: Mark-Houwink plot of high molecular weight precursor and samples after 10, 90 and 1440 minutes; $[E]_0/[I]_0 = 15$; (■): $[I]_0 = 1 \cdot 10^{-3}$ mol/l, (○): $[I]_0 = 2 \cdot 10^{-3}$ mol/l; (----): slope for linear PtBuA[18]

A maximum of the intrinsic viscosity has also been observed for dendrimers[19-21]. Here, the maximum occurs in the third generation, $G = 3$. At higher generations the intrinsic viscosity decreases with molecular weight. This is explained with a change in structure. After the addition of certain generations the dendrimer behaves like a sphere and can be described by Einstein's law: $[\eta] = 2.5 V_h/M$. While adding the new generation the increase of the hydrodynamic volume, $V_h \propto G^3$, is smaller than the increase of molecular weight, $\log M \propto G$ and therefore $[\eta]$ decreases. This is due to the fact that the density in the center of the dendrimer is much smaller than in the outer sphere. For the star polymers the density is also changing but here the density in the inner region is much higher than in the outer. After a certain number of arms are added the increase of V_h is only very small but molecular weight still increases linearly. Another difference between these star polymers and dendrimers is that the star polymers are not self-similar. The structure is not constant while arms are added, and therefore the obtained exponent α in the Mark-Houwink plot is not constant either. It cannot be used to draw conclusions on the shape of the polymer in solution.

Light scattering. The characterization of the star polymers with short arms (low molecular weight precursor) by GPC-MALLS was not possible. The scattering signal in THF is too small because of the small refractive index increment of PtBuA (dn/dc = 0.059 ml/g) and the low molecular weight. The stars with the longer arms were analyzed using additionally an online scanning interferometer to determine dn/dc. In analogy to the viscosity measurements, the relation between the radius of gyration and the molecular weight was investigated. The molecular weights determined by GPC-MALLS are summarized in Table 3 together with the exponents α and α_s, the exponent of the relation R_g vs. molecular weight. In Figure 9 the concentration signal and the radius of gyration vs. the molecular weight can be seen. There are reasonable results only for M > 300,000. The radius of gyration increases only slightly with the number of arms.

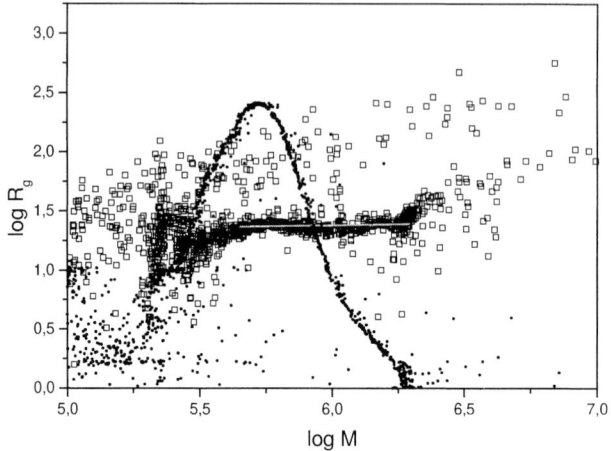

Figure 9: Radius of gyration (□) for a star polymer with long precursor; $[E]_0/[I]_0 = 15$; $[I]_0 = 1 \cdot 10^{-3}$ mol/l; (•): concentration signal; $\alpha_s = 0.038$.

The analysis is quite difficult due to the fact that there is no scattering signal for the residual precursor. This also poses problems in the determination of the molecular weight averages. M_n and M_w obtained from GPC-MALLS are always higher than those derived from GPC-viscosity using a universal calibration curve and determined for the whole sample including the residual precursor (Table 3). This results in a considerable error in M_n due to undetectable residual precursor. Nevertheless, the values for M_w are within 15 % error compared to GPC-viscosity indicating that GPC separation is working for these polymers, even when the previous results showed that the dimensions of the stars only slightly increase with molecular weight (i. e., number of arms). For reliable universal calibration the separation of the polymers is required because the method is only relative and a calibration curve and monodisperse slices are needed. GPC-MALLS is an absolute method and the obtained M_w values should be correct even if the GPC columns are not able to properly separate molecules with different arm numbers. Nevertheless the separation is needed for the correct determination of the exponent α_s. If the slices are polydisperse the weight-average of the molecular weight and the z-average of the radius of gyration are compared in every slice and this can cause an error in the evaluation[22].

The obtained exponents, α_s, are extremely low and smaller than for hard spheres with constant density ($\alpha_s = 0.33$). However, they are in the range calculated by Daoud and Cotton ($\alpha_s = 0.2$).

Table 3: Molecular weights and scaling exponents obtained from GPC-MALLS and GPC-viscosity coupling

$M_{w,Pre}$ ·10^{-4}	$[E]_0/[I]_0$	$[I]_0·10^3$, mol/l	t, min	$M_w·10^{-5}$ (GPC-MALLS)	deviation to M_w (GPC-visco)[a] %	α_S	α
10.2	5	1	90	3.04	9	-	0.129
10.9	10	1	90	4.19	13	0.140	0.032
10.9	15	1	90	5.03	12	0.038	0.036
8.90	15	2	90	5.27	7	0.198	-0.062
8.28	15	2	1440	10.6	12	0.233 0.787[b]	-0.099

[a] for the sample including the residual precursor, therefore M_w and especially M_n are lower than the average molecular weights given in Table 1 and Table 2
[b] high molecular weight peak

Figure 10 shows a plot of radius of gyration vs. molecular weight with two different slopes. For the "normal" star polymers a small exponent is observed ($\alpha_s = 0.233$). However, for M > 10^6 a much higher value is found ($\alpha_s = 0.787$). This may be due to a structural change. These polymers probably are coupled stars formed by core-core coupling which may be more ellipsoid than normal stars.

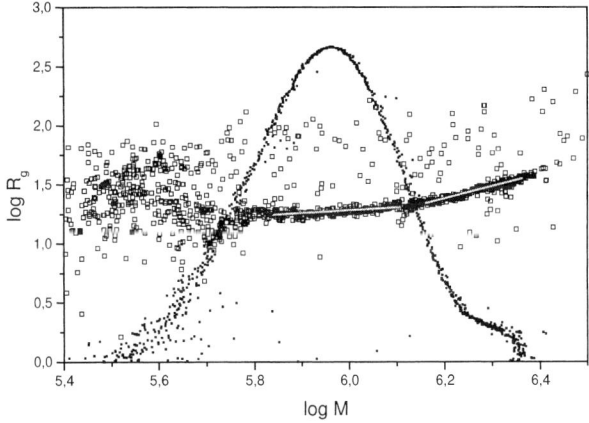

Figure 10: Determination of α_s for a sample with a high molecular weight precursor ($M_w = 72,000$) and star-star coupling; $[E]_0/[I]_0 = 15$; $[I]_0 = 2·10^{-3}$ mol/l

Computer Simulations

Computer simulations were performed in order to confirm the obtained low values for α_s. The radius of gyration was calculated for star polymers in a good solvent with constant number of segments per arm, N_{arm}, and different number of arms, f. From the plot of radius of gyration vs. total number of segments of the star polymer, N, (Figure 11) α_s can be calculated. The obtained values are very low and decrease with increasing number of segments per arm.

These results are in good agreement with the experimental data and indicates even smaller scaling factors than predicted by Daoud and Cotton.

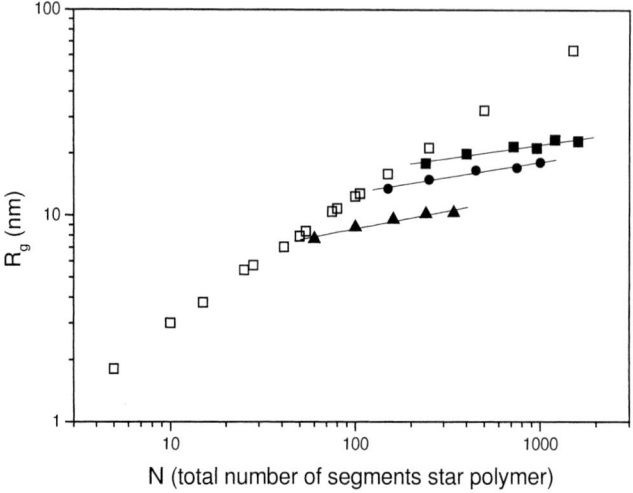

Figure 11: Results of computer simulation of star polymers with constant arm length, N_{arm}. ($\square$) linear polymer; $\alpha_s = 0.616$; ($\blacktriangle$) $N_{arm} = 20$; $f = 3\text{-}17$; $\alpha_s = 0.172$; ($\bullet$) $N_{arm} = 50$; $f = 3\text{-}20$; $\alpha_s = 0.147$; ($\blacksquare$) $N_{arm} = 80$; $f = 3\text{-}20$; $\alpha_s = 0.133$

Conclusions

For PtBuA star polymers with EGDMA core the precursor length, the ratio $[E]_0/[I]_0$ and the reaction time for the bifunctional monomer influence the average number of arms. This is in agreement with results of Mays, Burchard and Higashimura and their groups. The most important factor affecting the average number of arms is the overall concentration. Doubling of the concentrations resulted in star polymers with nearly doubled average number of arms. The amount of residual precursor decreases with decreasing precursor length, increasing time and increasing ratio $[EGDMA]_0/[I]_0$. All star polymers have a quite narrow arm number distribution. The viscosity behavior of the stars is similar to that of dendrimers. The intrinsic viscosity passes a maximum, it decreases after a certain number of arms is reached. In contrast to dendrimers the radius of gyration increases only slightly with molecular weight. This results in a low scaling exponent as predicted by Daoud and Cotton. The experimental data are in good agreement with simulations performed for star polymers in a good solvent.

Acknowledgement: This work was supported by the *Deutsche Forschungsgemeinschaft* within the *Schwerpunktsprogramm "Polyelektrolyte".* The authors thank Dr. Wolfgang Radke (DKI, Darmstadt) for helpful discussions.

References

1) S. Bywater, *Adv. Polym. Sci.* **30**, 89 (1979)

2) M. Morton, T. E. Helminiak, S. D. Gadkary, F. J. Bueche, *J Polym. Sci.* **57**, 471 (1962)

3) N. Hadjichristidis, *J. Polym. Sci., Part A.* **37**, 857 (1999)

4) H. Eschwey, M. L. Hallensleben, W. Burchard, *Makromol. Chem.* **173**, 235 (1973)

5) J.-G. Zilliox, P. Rempp, J. Parrod, *J Polym. Sci.: Part C* **22**, 145-156 (1968)

6) PCT Patent WO 86/00626 (1986), E. I. du Pont de Nemours & Co., inv. J. H. Spinelli

7) J. A. Simms, *Rubber Chemistry And Technology* **64**, 139 (1991)

8) C. Tsitsilianis, P. Lutz, S. Graff, J. P. Lamps, P. Rempp, *Macromolecules* **24**, 5897 (1991)

9) V. Efstratiadis, G. Tselikas, N. Hadjichristidis, J. Li, Y. Wan, J. W. Mays, *Polym. Int.* **33**, 171 (1994)

10) P. Lang, W. Burchard, M. S. Wolfe, H. J. Spinelli, L. Page, *Macromolecules* **24**, 1306 (1991)

11) S. Kanoka, M. Sawamoto, T. Higashimura, *Macromolecules* **24**, 2309 (1991)

12) D. M. Haddleton, M. C. Crossman, *Macromol. Chem. Phys.* **198**, 871 (1997)

13) M. C. Crossman, D. M. Haddleton, *Macromol. Symp.* **132**, 187 (1998)

14) J. Roovers, L.-L. Zhou, P. M. Toporowski, M. van der Zwan, H. Iatrou, N. Hadjichristidis, *Macromolecules* **26**, 4324 (1993)

15) M. Daoud, J. P. Cotton, *Journal de Physique* **43**, 531 (1982)

16) W. Burchard, *Adv. Polym. Sci.* **143**, 113 (1999)

17) W. Radke, Dissertation, Universität Mainz Mainz 1996

18) L. Mrkvicková, J. Danhelka, *J. Appl. Polym. Sci.* **41**, 1929 (1990)

19) O. Lambert, P. Dumas, G. Hurtrez, G. Riess, *Macromol. Rapid Commun.* **18**, 343-351 (1997)

20) C. Cai, Z. Y. Chen, *Macromolecules* **31**, 6393 (1998)

21) F. Bohlmann, J. Jacob, *Chem. Ber.* **107**, 2578 (1974)

22) M. Wintermantel, M. Gerle, K. Fischer, M. Schmidt, I. Wataoka, K. Urakawa, Y. Tsukahara, *Macromolecules* **29**, 978 (1996)

Synthesis and Morphology of Model 3-Miktoarm Star Terpolymers of Styrene, Isoprene and 2-Vinyl Pyridine

Agapi Zioga, Stella Sioula and Nikos Hadjichristidis*

Department of Chemistry, University of Athens, Panepistimiopolis,

Zografou, 157 71 Athens, Greece

SUMMARY: The combination of anionic polymerization and controlled chlorosilane chemistry made possible for the first time the synthesis of model 3-miktoarm star terpolymers of styrene (PS), isoprene (PI) and 2-vinylpyridine (P2VP) (3μ-SIV). The morphology of a nearly symmetric 3μ-SIV star terpolymer, was also studied. From the preliminary results, it seems that the PI and P2VP phases form hexagonally packed adjoined cylinders, whereas the PS phase occupies the remaining space forming non-regular curved hexagons, hexagonally packed as well. The star junction points reside on periodically spaced, parallel lines defined by the intersection of the three microdomain interfaces. Non of the phases form the matrix. The star molecular architecture gives the molecule the ability to "choose" which arms directly interact in the microphase segregate state, in order to minimize the most highly unfavorable contact between the PI and P2VP arms.

Introduction

Designing molecules with specific properties for certain applications has been of recent interest both scientifically and technologically. The key point in synthesizing polymeric materials with predetermined properties, and thus applications, is to know the relationship between structure and properties. In order to study this relationship, it is essential to synthesize model megamolecules with well defined macromolecular architecture and high degrees of molecular and compositional homogeneity.

Anionic polymerization, high vacuum techniques and controlled chlorosilane chemistry are very powerful tools for manipulating macromolecular architecture. Controlled substitution of chlorine atoms of chlorosilanes by macroanions produces a

© WILEY-VCH Verlag GmbH, D-69469 Weinheim, 2000   CCC 1022-1360/00/$ 17.50+.50/0

wide variety of model non-linear block copolymers with high degrees of molecular and compositional homogeneity and complex architectures [1].

Although non-linear block copolymers have been extensively studied, leading to very interesting results [2], rather little attention has been given to the non-linear architecture of terpolymers (3-miktoarm star terpolymers), due to synthetic difficulties. A number of synthetic approaches have been used for the synthesis of 3-miktoarm star terpolymers. The controlled chlorosilane chemistry was used in the case of the 3-miktoarm star terpolymers of PS, 1,4-PI and 1,4-polybutadiene (1,4-PBd) (3μ-SIB) [3]. The macromonomer approach was used for the synthesis of the 3-miktoarm star terpolymers of PS, poly(dimethylsiloxane) (PDMS) and poly(tert-butyl methacrylate) (PtBMA) [4], and of PS, PBd-1,2 and PMMA [5]. In the case of the 3-miktoarm star terpolymers of PS, 1,4-PI and poly(methyl methacrylate) (PMMA) (3μ-SIM), incorporation of the polymethacrylate branch by the chlorosilane method has not been possible, simply because the reaction of SiCl with the living polymethacrylate fails to give the linked product. In order to overcome this difficulty, the SiCl group was transformed to $SiC(Ph_2)(CH_2)_2C^-(Ph_2)Li^+$, which is an efficient initiator for the polymerization of the methyl methacrylate (MMA) [6]. Finally, the synthesis of 3-miktoarm star terpolymers of PS, poly(ethylene oxide) and poly(ε-caprolactone) was performed by preparing first a diblock copolymer having a protected anionic initiator at the junction point, followed by deprotection and polymerization of the third compound [7].

The challenging synthesis of suitable model star polymers makes quite difficult the morphological study of such materials and the construction of a phase diagram. Morphological studies of the 3μ-SIB showed that the two dienes are mixed because of their low interaction parameter and the structure consists of only two kind of microdomains [8]. The nearly symmetric 3-miktoarm star terpolymers of PS, PDMS and PtBMA [9] seems to microphase separate into three microdomain structures, but no specific model was proposed. Finally in the case of the 3μ-SIM star terpolymers the microdomain structure was governed by the pronounced incompatibility of the PMMA and PI arms and the low interaction parameter between PS and PMMA. Thus, in the case of the nearly symmetric 3μ-SIM star terpolymers, the PI and PS phases form prism-shape columns (triangular and rectangular) surrounding a cylindrical core region of PMMA [10]. In these samples the junction points reside on periodically spaced, parallel lines defined by the intersection of the three microdomain interfaces. The non-symmetric (in respect of their composition) 3μ-SIM star terpolymers exhibited an inner PI column with a surrounding PS annulus in a matrix of PMMA. Depending on the composition and molecular weight PI/PS and PS/PMMA interfaces were either cylindrical (%vol. PS/PI/PMMA: 20.5/25/54.5, 22.3/27.2/50.5) or had a non-constant mean curvature (non-CMC) diamond prism shape (%vol. PS/PI/PMMA: 25.3/30.7/44, 29/49.3/21.7). The star junction points are distributed over the PI/PS intermaterial dividing surface due to partial mixing of PS and PMMA arms [11].

In all the above cases, all the three arms of the star terpolymers are amorphous. However, in the case of the 3-miktoarm star terpolymer of PS, poly(ethylene oxide) (PEO) and poly(ε-caprolactone) (PεCL) the situation is totally different. The megamolecule consists of two crystallizable arms (PEO and PεCL) and one amorphous (PS). Thus, crystallization is the main force that drives the microphase separation and the final structure formation [12].

In this paper we report the synthesis and preliminary results of the morphological characterization of 3-miktoarm star terpolymers of styrene, isoprene and 2-vinyl pyridine, in order to study the influence of the macromolecular architecture and the chemical nature of the arms on the properties of these polymeric materials.

Experimental

The synthesis of 3μ-SIV was performed using anionic polymerization, high vacuum techniques and controlled chlorosilane chemistry. The first step is the synthesis of the macromolecular monofunctional chlorosilane linking agent $(PS)(PI)(CH_3)SiCl$ according to the standard procedure [3], followed by condensation of the benzene solution and dilution with tetrahydrofuran (THF). In order to avoid side reactions, the polymerization of 2-vinyl pyridine was performed in THF at $-78°C$ using 1,1-diphenylhexyllithium as initiator. The solution of $(PS)(PI)(CH_3)SiCl$ was added to an excess of P2VP$^{(-)}$Li$^{(+)}$ in THF at $-78°C$, in order to incorporate the third arm. The excess of P2VP was extracted with treatment of the final product with methanol.

All steps were monitored by size exclusion chromatography (SEC). The number average weight, M_n, was determined in toluene by membrane osmometry (MO) at $35°C$ with a Wescan 230 membrane osmometer. The weight average molecular weight, M_w, was measured with a Chromatix KMX-6 , operating at $\lambda=633nm$ in THF at $25°C$. The refractive index increment (dn/dc) was determined with a Chromatix KMX-16 operating at $\lambda=633nm$.

The composition of the terpolymers was analyzed by ^{1}H-NMR in CDCl$_3$ using a Vanity Unity Plus 200 instrument. The PI arm analyzed by ^{1}H and ^{13}C-NMR was found to have the following microstructure : 70% wt cis 1,4; 20% wt trans 1,4 and 10%wt 3,4.

For Small Angle X-Ray Scaterring (SAXS) and Transmission Electron Microscopy (TEM) characterization the nearly symmetric 3μ-SIV sample was slowly cast into a ~1 mm thick film from THF solution (~4 wt %) and annealed at $135°C$ under vacuum for 10 days in order to obtain near-equilibrium morphology. For TEM investigation 40-50nm thick sections were cryomicrotomed at $-120°C$ using a Reichert-Jung FC 4E cryoultramicrotome equipped with a diamond knife. Sections were picked on copper-grids and selectively stained with OsO$_4$ or CH$_3$I. The morphology was observed with a Jeol 200CX transmission electron microscope operating at an accelerating voltage of 200kV in the bright field mode.

Results and Discussion

Synthesis

The synthesis of the 3-miktoarm star terpolymers of polystyrene (PS), polyisoprene (PI) and poly(2-vinyl pyridine) (P2VP), 3μ-SIV, was performed according to the following basic reactions:

Isoprene + s-BuLi $\xrightarrow{C_6H_6}$ PI$^-$Li$^+$

PI$^-$Li$^+$ + excess (CH$_3$)SiCl$_3$ $\longrightarrow$ PI(CH$_3$)SiCl$_2$ + LiCl + (CH$_3$)SiCl$_3$ $\uparrow$

Styrene + s-BuLi $\xrightarrow{C_6H_6}$ PS$^-$Li$^+$

PI(CH$_3$)SiCl$_2$ + PS$^-$Li$^+$ $\xrightarrow{\text{titration}}$ (PI)(PS)(CH$_3$)SiCl + LiCl

s-BuLi + CH$_2$=C(Ph)$_2$ $\xrightarrow[-78^0C]{\text{THF}}$ s-Bu - CH$_2$ - C$^-$(Ph)$_2$Li$^+$

s-Bu - CH$_2$ - C$^-$(Ph)$_2$Li$^+$ + 2-Vinylpyridine $\xrightarrow[-78^0C]{\text{THF}}$ P2VP$^-$Li$^+$

P2VP$^-$Li$^+$ (excess) + (PI)(PS)(CH$_3$)SiCl $\xrightarrow[-78^0C]{\text{THF}}$ (PS)(PI)(P2VP)   [3μ-SIV]

The monofunctional linking agent, (PS)(PI)(CH$_3$)SiCl, was synthesized by controlled substitution of two chlorine atoms of trichloromethylsilane by PS and PI arms, using methods developed by our group [3]. Polyisoprenyllithium, prepared in benzene with sec-butyllithium (s-BuLi) as initiator, was reacted with a large excess of trichloromethylsilane (TCMS), in order to substitute only one of the three chlorine atoms with the PI arm (Fig. 1). After careful removal of the TCMS, the (PI)Si(CH$_3$)Cl$_2$ was dissolved in benzene and titrated with polystyryllithium, prepared in benzene with s-BuLi. The titration was monitor by SEC (Fig. 1). Subsequently the solution of (PS)(PI)(CH$_3$)SiCl was condensed and diluted with THF. The incorporation of the P2VP arm was achieved by reacting the (PS)(PI)(CH$_3$)SiCl with an excess of the living anion P2VP$^-$Li$^+$, prepared in THF at −78°C using 1,1-diphenylhexyllithium as initiator. This last reaction was performed in polar solvent (THF) and at low temperatures (-78°C), in order to avoid termination reactions of the P2VP$^-$Li$^+$ anion. In order to eliminate the excess of the P2VP arm, the reaction product was treated with methanol (Fig. 1), which is a good solvent for P2VP and a non-solvent for the 3μ-SIV.

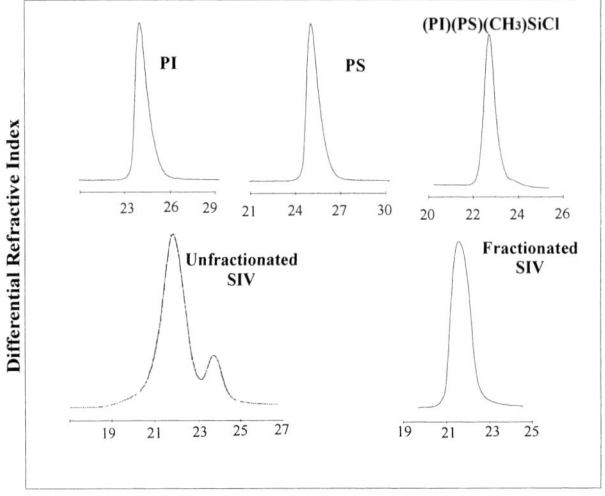

Elution Volume (mls)

Fig.1: Size exclusion chromatograms for monitoring the synthesis of 3-miktoarm star terpolymer SIV-15/14/13.

The molecular characteristics of the fractionated star terpolymers are given in Table 1. The nomenclature used is SIV-$x/y/z$, where x, y, z are the respective arm molecular masses in kg/mol.

Table 1. Molecular characteristics of the 3-miktoarm star terpolymers

Polymer	$M_w \cdot 10^{-3}$ [a]	$M_n \cdot 10^{-3}$ [b]	M_w/M_n [c]	PS/PI/P2VP (%w.f.) [d]
PS	15	14.5	1.02	-
PI	14.5	14	1.02	-
SIV-15/14/13	45		1.1	34.2/34/31.8
PS	35	35	1.02	-
PI	15.5	15	1.02	-
SIV-35/15/20	70		1.09	57.7/22.3/20

[a] LALLS in THF at 25^{0}C
[b] MO in toluene at 35^{0}C
[c] SEC in THF at 30^{0}C
[d] ^{1}H-NMR in CDCl$_3$ at 35^{0}C

The above results indicate a high degree of molecular and compositional homogeneity.

Morphological Characterization

The preliminary results of the nearly symmetric 3-miktoarm star terpolymer, SIV-15/14/13, indicate a novel, very interesting morphological behavior for this kind of materials. The degree of segregation $\chi_{ij}(N_i+N_j)$ and the lowest order d spacing of SIV-15/14/13 are given in Table 2. The Flory-Huggins interaction parameters of PS-PI and PS-P2VP at 135^0C are determined as ~0.05 and 0.12 respectively [13-14]. Based on the solubility parameters [THF: δ = 9.1 (cal/cm^3)$^{1/2}$, PI: δ = 8.2 (cal/cm^3)$^{1/2}$, PS: δ = 9.1 (cal/cm^3)$^{1/2}$, P2VP: δ = 10 (cal/cm^3)$^{1/2}$] the χ parameter between PI and P2VP is expected to be the highest, $[\chi_{ij} \sim (\delta_i - \delta_j)^2]$.

Table 2. The lowest order d spacing and the degree of segregation of SIV-15/14/13

Polymer	PS/PI/P2VP (%v.f.)[a]	ϕ_1: ϕ_2: ϕ_3	$\chi_{SI}N_{SI}$[b]	$\chi_{SV}N_{SV}$[b]	d_1^{SAXS}(Å)
SIV-15/14/13	33.7/38.1/28.2	1.2/.1.3/1	18	32	249

[a] From %w.f. from ^{1}H-NMR and ρ(PS)=1.06 g/ml, ρ(PI)=0.913 g/ml and ρ(P2VP)=1.1 g/ml
[b] $\chi_{ij}(N_i+N_j)$ at 135^0C.

The results from SAXS and TEM for the THF cast SIV-15/14/13 sample are given in Fig.2 and 3, respectively. The SAXS pattern suggests that the sample exhibits a hexagonal arrangement of microdomains, because the q_n : q_1 values are 1.0: 1.7: 1.96: 2.61, very close to 1: $\sqrt{3}$: $\sqrt{4}$: $\sqrt{7}$, which are characteristic for hexagonally packed units.

The bright field TEM micrograph of section stained with OsO$_4$ is given in Fig. 3a. OsO$_4$ preferentially stains the PI phase, which appears dark, whereas the PS and P2VP phases are not distinguished and appear as the light matrix. From this micrograph it is obvious that the PI phase forms hexagonally packed cylinders in a matrix of PS and P2VP. The TEM micrograph of section stained with CH$_3$I, which stains preferentially the P2VP phase, shows that the P2VP microdomains, which appear dark, also form hexagonal cylinders in a matrix of PS and PI (Fig. 3b.).

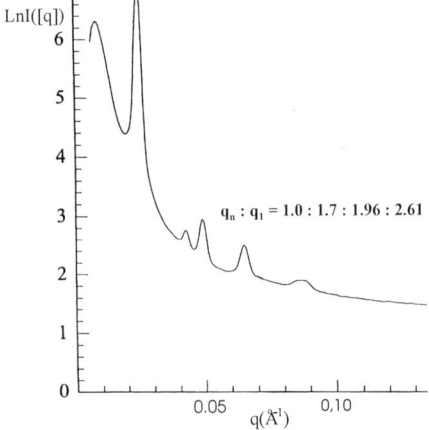

Fig. 2: SAXS pattern of SIV-15/14/13. The ratios of the q vectors for the various peaks are also given.

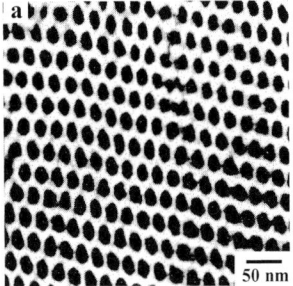 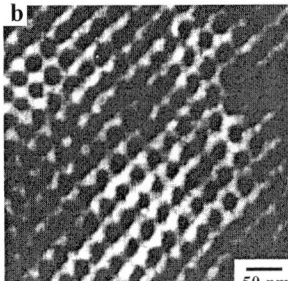

Fig. 3: Bright field TEM images of SIV-15/14/13 stained with (a) OsO_4 and (b) CH_3I. In the OsO_4 stained micrograph, the dark regions correspond to the PI phase, which forms hexagonally packed cylinders, whereas in the CH_3I stained micrograph the dark regions correspond to the P2VP phase, which also seems to form hexagonally packed cylinders.

Combining the two sets of TEM data with the hexagonal packing indicated by SAXS and the star macromolecular architecture leads to the schematic given in Fig. 4 for the proposed morphology. The PI and P2VP phases form hexagonally packed cylinders, whereas the PS phase forms non-regular curved hexagons, hexagonally packed as well. The two kinds of cylinders adjoin at the star junction points, which thus reside on periodically spaced, parallel lines defined by the intersection of the three microdomain interfaces. The structure is two-dimensionally periodic with plane group symmetry p3m1. All components form one-dimensionally continuous domains and there is no matrix component.

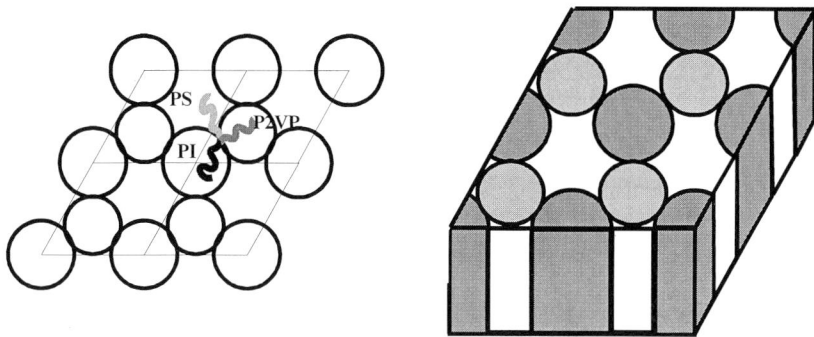

Fig. 4: Schematic representation of the proposed microdomain morphology of SIV-15/14/13 sample. The chain conformation of the three blocks and the location of the junction point are also presented.

Comparing the preliminary morphological results for the SIV-15/14/13 sample with the microdomain structure of the nearly symmetric SIM 3-miktoarm star terpolymers[10] (ϕ_1/ ϕ_2/ ϕ_3: SIV-15/14/13 = 1.2/1.3/1, SIM-72/77/109 = 1/1.2/1.3 and SIM-92/60/94 = 1.3/1/1.2) a lot of similarities are observed. First of all, there is no interface between the most incompatible phases, that is the PI/P2VP and the PI/PMMA phases. The star molecular architecture gives the molecule the ability to "choose" which arms directly interact in the microphase segregate state. Since χ_{IV} and χ_{IM} have the highest value, the arms microphase separate to obtain microdomain structures which minimize the unfavorable contact between the PI/P2VP and PI/PMMA arms. In addition the star junction points reside on periodically spaced, parallel lines defined by the intersection of the three microdomain interfaces, while non of the phases form the matrix.

A possible scenario for the formation pathway of the proposed structure is as follows: THF was the solvent used for casting. According to the solubility parameters of the solvent and the three components, it is most probable that PI and P2VP segregate first forming, according to their volume fraction (38% and 28% respectively), cylinders hexagonally packed in a swollen matrix of PS. Since these two blocks are covalently bonded to each other, these two kind of cylinders are obliged to adjoin at the junction point of the star. That way the interface between the two most incompatible arms (PI and P2VP) is minimum. Upon further evaporation of the solvent the PS phase occupies the remaining space forming this strange shape column, while there is no need for any deformation of the two kind of cylinders, since there interface is already minimum.

In order to confirm the proposed morphology and the similarities observed with the nearly symmetric SIM 3-miktoamr star terpolymers, more detailed examination of the sample is needed. The next step is to perform a TEM experiment using two kind of stainers (e.g. both OsO_4 and CH_3I) in order to achieve two levels of contrast and be able to "see" the two kind of cylinders in the same micrograph [15].

Conclusions

Controlled chlorosilane chemistry can be successfully used for the synthesis of 3-miktoarm star terpolymers of styrene, isoprene and 2-vinyl pyridine. It is a general and very powerful method for the synthesis of model megamolecules with complex architecture, that was up to now mainly used with non-polar monomers (such as styrenic and dienic) and now expanded to the polar 2-vinyl pyridine monomer, opening new horizons for the synthesis of model polymeric materials with more potential applications. The preliminary study of the morphology of the nearly symmetric 3μ-SIV sample revealed a very interesting microstructure, consisting of two kind of hexagonally packed cylinders of PI and P2VP adjoining at the star junction points, which reside on periodically spaced parallel lines defined by the intersection of the three microdomains. Very interesting results can also be obtained by comparing the nearly symmetric 3μ-SIV with the corresponding 3μ-SIM star terpolymers. Although the chemical nature of the third arm of the two kinds of 3-

miktoarm star terpolymers is different (P2VP versus PMMA), some common rules are followed for the formation of the microdomain structure: minimization of the most unfavorable contact and localization of the star junction points on periodically spaced parallel lines. A more detailed future study of the 3μ-SIV star terpolymers with various compositions is necessary in order to propose a phase diagram for this system and to further use it for predicting the morphological behavior and thus the properties of this kind of polymeric materials[15].

Acknowledgements

This work was conducted in collaboration with Prof. E. L Thomas (Department of Material Science and Engineering, MIT, MA).

References

1. N. Hadjichristidis, *J. Polym. Sci., A: Polym. Chem.* **37**, 857 (1999).
2. D. J. Lohse and N. Hadjichristidis, *Current Opinion in Colloid & Interface Science* **2**, 171 (1997).
3. H. Iatrou, N. Hadjichristidis, *Macromolecules* **25**, 4649 (1992).
4. T. Fujimoto, H. Zhang, T. Kazama, Y. Isono, H. Hasegawa, T. Hashimoto, *Polymer* **33**, 2208 (1992).
5. H. Hückstädt, V. Abetz, R. Stadler, *Macromol. Rapid Commun.* **17**, 599 (1997).
6. S. Sioula, Y. Tselikas, and N. Hadjichristidis, *Macromolecules* **30**, 1518 (1997).
7. O. Lambert, P. Dumas, G. Hurtez, G. Reiss, *Macromol. Rapid Commun.* **18**, 343 (1998).
8. N. Hadjichristidis, H. Iatrou, S. Dehal, J. J. Chludzinski, M. Disko, R. T. Garner, K. S. Liang, D. J. Lohse, S. T Milner, *Macromolecules* **26**, 5812 (1993).
9. S. Okamoto, H. Hasegawa, T. Hashimoto,T. Fujimoto, H. Zhang, T. Kazama, A. Takano, Y. Isono *Polymer* **38**, 5275 (1997).
10. S. Sioula, N. Hadjichristidis, E.L. Thomas, *Macromolecules*, **31**, 8429 (1998).
11. S. Sioula, N. Hadjichristidis, E.L. Thomas, *Macromolecules*, **31**, 5272 (1998).
12. G. Floudas, G. Reiter, O. Lambert, P. Dumas, *Macromolecules* **31**, 7279 (1998).
13. N. P. Balsara, Thermodynamics of Polymer Blends. In *Physical Properties of Polymer Handbook*, J. E. Mark, Ed. AIP: New York, 1996; Chapter 19.
14. K. H. Dai, E. J. Kramer, *Polymer* **35**, 157 (1994).
15. A. Zioga, S. Sioula, N. Hadjichristidis, E. L. Thomas, to be published.

Macromol. Symp. **157**, 251–257 (2000)

TITANIUM AND VANADIUM BASED NON-METALLOCENE CATALYSTS FOR OLEFIN POLYMERIZATION

Kazuo Takaoki and Tatsuya Miyatake[*]

Petrochemicals Research Laboratory, Sumitomo Chemical Co. Ltd.
2-1 Kitasode, Sodegaura, Chiba 299-0295, Japan

Abstract: Titanium complexes with chelating 2,2'-thiobis(4-methyl-6-R-phenoxy) ligand (R= CH_3, i-C_3H_7, t-C_4H_9, $Si(i$-$C_3H_7)_3$) ligands in combination with methylalumoxane, methylisobutylalumoxane or triisobutyl-aluminum/$(C_6H_5)_3CB(C_6F_5)_4$ as cocatalyst were highly active toward olefin, giving polymers with high molecular weight. Especially, titanium complexes having the most bulky substituent of $Si(i$-$C_3H_7)_3$ showed the highest activity of ethylene polymerization. Vanadyl complexes with 2,2'-thiobis(4-methyl-6-$tert$-butylphenoxy) ligand prepared by the reaction with $VO(OC_4H_9)_3$ were also highly active to propylene, giving isotactic polypropylene with the melting temperature of 138°C.

INTRODUCTION

Titanocenes and titanium tetraalkoxides combined with an alkylaluminum compound can polymerize olefins.[1] However, the catalytic activities are very low compared with that of commercial catalysts, i.e., $TiCl_3$, and $MgCl_2$-supported catalyst. On the other hand, Kaminsky and Sinn found that zirconocene compounds, when they are combined with methylalumoxane (MAO) give polyethylene in excellent catalytic activiy.[2] We have focused on the application of MAO to a non-metallocene titanium complex catalyst system such as titanium alkoxide or titanium phenoxide and have found the novel titanium complex catalyst with 2,2'-thiobis(4-methyl-6-$tert$-butylphenoxy) ligand (**1**, (TBP)$TiCl_2$) showing high activity to the polymerization of propylene as well as ethylene.[3] This catalyst system is considered to be promising system not only for a displacement of zirconocene catalyst system as a single-site olefin polymerization catalyst system but also for synthesis of new polymer materials. (TBP)TiCl₂/MAO system is active for olefins, styrene, conjugated dienes, and non-conjugated dienes.[4,5] The catalyst performance of (TBP)TiCl₂/MAO system is summarized in Scheme 1. This catalyst system gives (1) amorphous polypropylene with extremely high molecular weigh of more than 8 million, (2) polybutadiene with high *cis*-1,4-structure, (3) poly(1,5-hexadine) with complicate cyclic structure, (4) syndiotactic polystyrene, and (5) isotactic poly(ethylene-*alt*-styrene).

Scheme 1

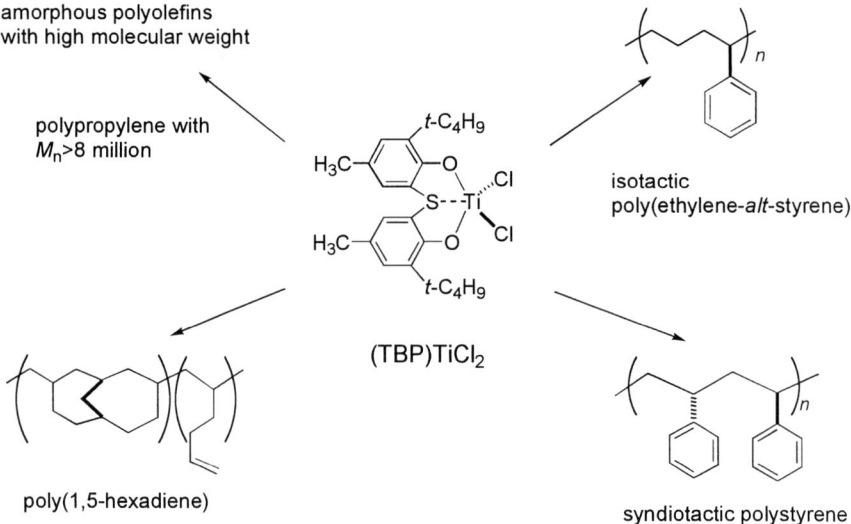

amorphous polyolefins
with high molecular weight

polypropylene with
M_n>8 million

isotactic
poly(ethylene-*alt*-styrene)

(TBP)TiCl₂

poly(1,5-hexadiene)

syndiotactic polystyrene

In a present paper, we report on the results of ethylene polymerization with new thiobis(phenoxy) titanium complex catalysts (**2-5**) and of propylene polymerization with vanadyl complex catalyst with 2,2'-thiobis(4-methyl-6-*tert*-butylphenoxy) ligand (**6** and **7**).

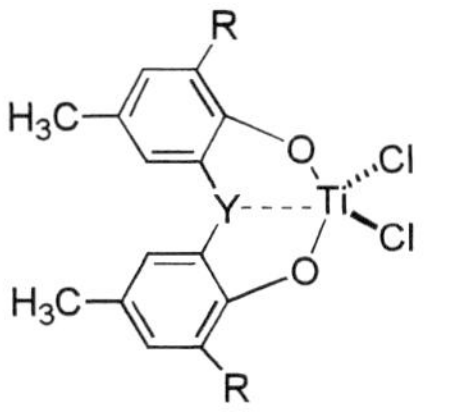

1; R = *t*-C$_4$H$_9$, Y = S
2; R = CH$_3$,   Y = S
3; R = *i*-C$_3$H$_7$, Y = S
4; R = Si(*i*-C$_3$H$_7$)$_3$, Y = S
5; R = *t*-C$_4$H$_9$, Y = SO$_2$

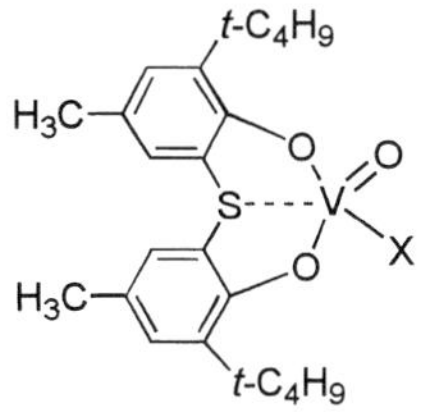

6; X = Cl, (TBP)VOCl
7; X = OC$_4$H$_9$, (TBP)VO(OBu)

RESULTS AND DISCUSSION

Synthesis of catalysts: Bulky phenols were synthesized according to the procedures shown in Scheme 2.   Triisopropylsilyl-substituted cresol was prepared from corresponding bromide.   After O-silylation of cresol, the silyl group was transferred to the 2-position of cresol with *tert*-butyllithium. Isopropyl-substituted cresol was regiospecifically prepared from *p*-cresol by 4 steps.   After Fries-rearrangement of cresol ester, acyl group was converted to isopropyl group by 2 steps.

Scheme 2

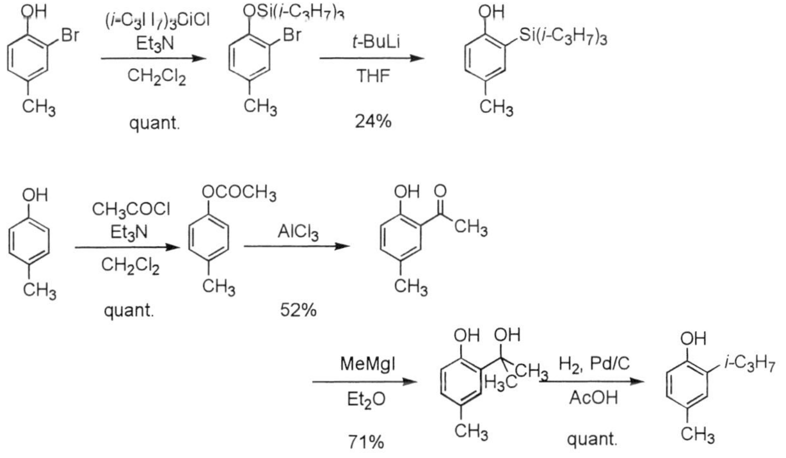

Preparation procedures of titanium complexes were summarized in Scheme 3. Thiobisphenols were prepared from corresponding phenols, after treatment of 2-substituted cresols with sulfer dichloride, resulted bis-phenol ligands were complexed with titanium tetrachloride in good yield.

The sulfonylene bridged bisphenol was prepared through oxidation of the sulfanylene bridged bisphenol, complexation of the bisphenol with titanium tetrachloride gave desired titanium complex in good yield.

Scheme 3

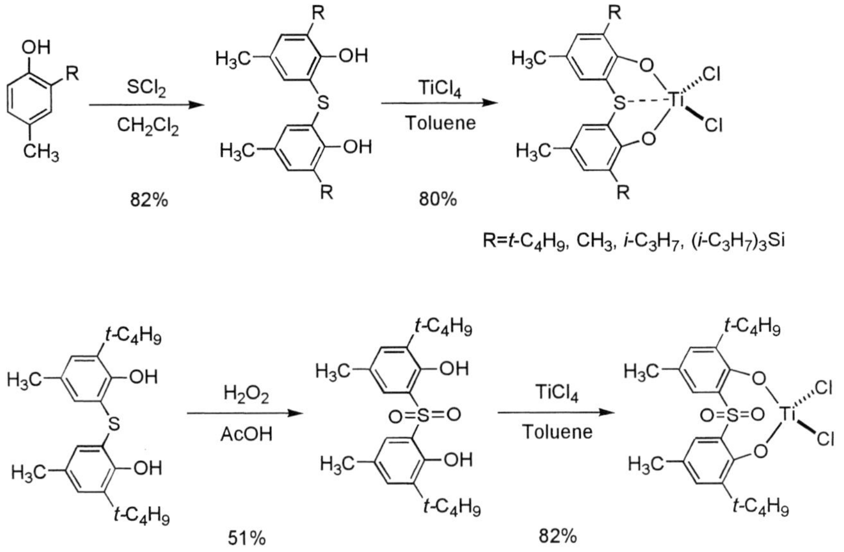

$R = t\text{-}C_4H_9,\ CH_3,\ i\text{-}C_3H_7,\ (i\text{-}C_3H_7)_3Si$

Vanadyl complex **6** was prepared by the reaction of 2,2'-thiobis(4-methyl-6-*tert*-butylphenol) with $VOCl_3$ in pentane at 0°C for 5 h in a 70% yield. Complex **7** was prepared by the reaction with $VO(OC_4H_9)_3$ in *n*-pentane at 20°C for 20 h in a 75% yield.

Polymerization of ethylene with titanium-based catalyst systems **1-5**: The results of ethylene polymerization with complexes **1-5** were summarized in Table 1. Three types of cocatalysts were used in this polymerization;

methylalumoxane (MAO), methyl-isobutylalumoxane (MMAO), and trityl-tetrakis(pentafluoro)borate/triisobutylaluminum system.   In the cocatalyst systems of MAO and borate, toluene was used as a solvent, and in the case of MMAO, n-heptane was used.   Polymerization conditions are shown in the footnote of the table.

Table 1. Polymerization of ethylene with various titanium complexes

complex	cocatalyst					
	MAO[a]		MMAO[b]		$(C_6H_5)_3CB(C_6F_5)_4/Al(^iC_4H_9)_3$[c]	
	activity (kg/mol Ti h)	$10^{-4} \times M_w$	activity (kg/mol Ti h)	$10^{-4} \times M_w$	activity (kg/mol Ti h)	$10^{-4} \times M_w$
2	170	—	1070	14	40	—
3	570	—	1900	17	190	—
1	550	27	3090	44	210	24
4	1050	71	670	73	880	154
5	40	48	780	22	90	51

a) polymerization conditions: 1 L autoclave; solvent = toluene, 300 mL; MAO/Ti = 1000 (mol/mol); 60°C for 1 h; ethylene = 4 atm.
b) MMAO/Ti = 1000; solvent = n-heptane, 300 mL.
c) Al/Ti = 500, Borate/Ti = 1; solvent = toluene, 300 mL.

Complex **2** having smallest substituent of methyl group among them showed the lowest activity in all cocatalyst systems.   On the other hand, complex **4** which has largest substituent of triisopropylsilyl group showed the highest activity and the polymer obtained had the highest molecular weight.

Sulfonylene bridged complex **5** was not so active toward ethylene polymerization compared with sulfanylene bridged complexes **1-4**.

These results indicate that a precise designing of a ligand structure and a suitable combination of complex and cocatalyst might improve a catalyst performance of such phenoxy-titanium complex catalyst system.

Polymerization of propylene with vanadium-based catalyst systems **6** and **7**: The results of propylene polymerization were summarized in Table 2. Both VOCl$_3$ and VO(OC$_4$H$_9$)$_3$ were active when combined with MAO. However, the molecular weight distribution of polymers obtained were very broad (M_w/M_n = 61 and 78). Both complexes **6** and **7** showed an excellent activity, which is comparable to that of Ti complex catalyst. By a introduction of thiobisphenoxy ligand to vanadyl compound, polypropylene with a narrow molecular weight distribution was obtained. Interestingly, polypropylene prepared with buthoxyvanadyl complex was isotactic-rich polymer (mm% is more than 35%). Especially, complex **7**/ MAO system gave a polypropylene with highest mm dyad tacticity of 68% and with the melting temperature of 138°C. This melting temperature is comparable to that of polypropylene prepared with ethylenebis(indenyl)zirconium dichloride/MAO system. Another characteristic of the polypropylene prepared with such vanadyl complex is that the polymer obtained has higher regioregularity (represented by F$_{000}$ + F$_{111}$) compared with (TBP)TiCl$_2$ (complex **1**) /MAO system[3].

Table 2. Polymerization of propylene with vanadyl complexes and MAO

complex	activity[a] (kg/mol M h)	$10^{-4} \times M_w$	M_w/M_n	stereoregularity [mm]%	Tm[b] (C)	regioregurarity[c] F$_{000}$ + F$_{111}$
(TBP)VOCl (**6**)	1900	190	3.2	35	—	0.98
(TBP)VO(OC$_4$H$_9$) (**7**)	420	220	4.0	68	138	—
(TBP)TiCl$_2$ (**1**)	8900	>800	2.2	22	—	0.54
VOCl$_3$	12	90	61	47	—	0.96
VO(OC$_4$H$_9$)$_3$	6	120	78	60	—	—

a) polymerization condition: cat., 1.1 x 10^{-3}mmol; MAO, 5.17 mmol; propylene, 30g; toluene, 3 mL; polymn. temp., 20°C; Time 1 h.
b) melting temperature   c) Y. Doi, *Macromolecules*, **12**, *248-251*, (1979)

The isotactic-rich polypropylene obtained with vanadyl complex **7** was a mixture of isotactic polypropylene and atactic polypropylene. The ^{13}C NMR spectra of methyl region of polypropylene fractions fractionated by an extraction with a boiling *n*-pentane are shown in Fig. 1. Spectrum (A) shows that the main defect of isotactic sequence is mmmmrrmmmm, which suggests that the isospecific propagation of propylene was governed by a catalytic control mechanism.

A more detailed designing of ligand and a selection of cocatalyst are necessary to improve the catalytic performance and to control the molecular structure of polymer obtained.

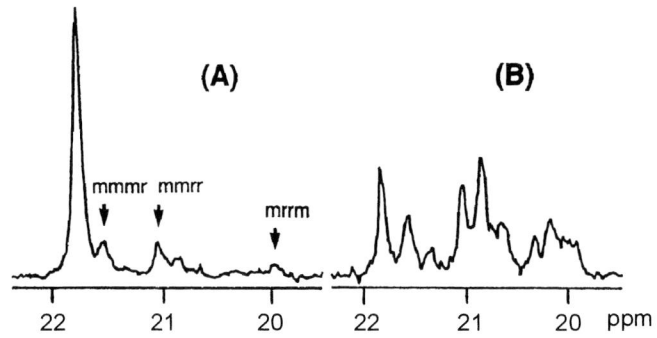

Fig. 1 ^{13}C NMR of PP fractions   (sample no. 2)
(A) fraction insoluble in *n*-pentane (71%)
(B) fraction soluble in *n*-pentane (29%)

REFERENCES

1) R.Fold, P.L. Cowe, in "The Organic Chemistry of Titanium", Butterworths, London 1965, pp 158-163.

2) For example, K.H. Brintzinger, D. Fischer, R. Mülhaupt, B. Rieger, and R.M. Waymouth, Angew. Chem. Int. Ed. Engl., 34, 1143 (1995).

3) T. Miyatake, K mizunuma, Y. Seki, and M. Kakugo, Makromol. Chem., Rapid Commun., 10, 349 (1989).

4) M. Kakugo, T. Miyatake, and K. Mizunuma, in "catalytic Olefin Polymerization", Kodansha Tokyo 1990, pp 517-529.

5) T. Miyatake, K. Mizunuma, and M. Kakugo, Makromol. Chem. Macromol. Symp., 66, 203 (1993).